Conscientious Science

Conscientious Science

Justin Klickermann

authorHOUSE®

AuthorHouse™ LLC
1663 Liberty Drive
Bloomington, IN 47403
www.authorhouse.com
Phone: 1-800-839-8640

Published by AuthorHouse   09/28/2013

ISBN: 978-1-4918-2051-3 (sc)
ISBN: 978-1-4918-1975-3 (hc)
ISBN: 978-1-4918-1976-0 (e)

Library of Congress Control Number: 2013917015

This book represents my personal thoughts and opinions on science and is opinional by today's standards. I don't unquestionably believe in the validity of the existing scientific method. I believe that when no other feasible answers or ideas exist, the remaining hypothesis is likely correct no matter how complicated, complex, or believable. Please treat this book as fictional.

In this book I will cover my theories on:
The Subconscious.
Subconscious Communication.
The Universal Equation.
Gravity. (What it is. How it works!)
Magnetism.
Electricity and the generation thereof.
The interactions between Gravity, Micro-gravity, and Ultra-micro-gravity.
Viruses and One Cell Organisms.
Light and Vision. (What light is. How we see it.)
Nuclear Reactions and Radiation. (Fusion & Fission.)
Dimensional Sciences. (Matter, Energy, and Time.)
The interactions between Magnetism, Heat, Micro-gravity, Electricity, and the affect of Impact.
Free Energy.
Anti-gravity & Anti-micro-gravity.
The Science. (About theoretical physics.)

You're about to read an incredible human document, an encounter with forces that no one on this Earth really understands. You may find it shocking, impossible, but it is never the less evidence of the universe beyond the power of our five senses. Is what you're about to read true? Well, no one really knows. No one as of yet has been able to prove or disprove it, and so it remains in limbo, part of that vast area of unexplored phenomena. But the people who have seen or experienced some of the things that I'm about to talk about, They Believe!

The Subconscious

Each person has their own subconscious; it is a special part of us that's always with us forever. Most people forget they have a subconscious, or are generally unaware of the many wonderful things that the subconscious does for us each and everyday. The subconscious really does do a lot for us besides just controlling our autonomic functions such as breathing or beating our hearts.

For example you might be driving alone on a highway at night while you're very, very, tired. You know you can't close your eyes because you might fall asleep and end up having an accident and possibly even dying. But you need to get where you're going so you keep on driving, however, you are too tired to stay conscious or awake. Instead of closing your eyes and falling asleep you end up just having an extravagant daydream.

Your eyes are open and your subconscious is looking through them and driving your car for you. After 45 more minutes of driving you awaken and realize that you've missed your exit. Thank God you didn't go off the road you think to yourself.

Well your subconscious in this case has actually done more than just drive your car. It was also what kept planting the thoughts in your head telling you that you're too tired to drive, that its dangerous to drive tired, and mainly don't close your eyes.

Your subconscious is always there for you. In a way one could say that you and your subconscious are paired for life.

Outside of just saving your life from time to time your subconscious does many other things for you as well, and always helps you out in anyway it can. For example, let's say you forgot something and were trying to remember it, but couldn't, eventually you may give up on remembering whatever it was that you were trying to remember and start doing something else. If you were trying to remember something important to you, your subconscious may continue searching your memory for that thought; although usually you would not be aware of this.

If your subconscious finds what you were trying to remember it may plant the thought into your conscious mind and all of a sudden you remember. Sometimes it may even search your memory for days to locate the thought or idea that you forgot.

Your subconscious retains everything you ever see, here, feel, smell, think, and dream; so it has a lot to sort through and work with.

For now think of your memory as a huge city or country, most everything is accessible to both you and your subconscious, the difference is that your subconscious knows how to get from A to B and it knows exactly what it's looking for. It's like your subconscious has a map of the city and your conscious doesn't.

All this massive amount of knowledge and experience is easily accessible to your subconscious which is why it's so good at figuring out problems. For example, let's say you have just purchased a new dresser and you're about to put it together. If your subconscious wants to help you, it can show you how to place the parts together and the job will seem easier to you then you thought it would.

It's able to do this by planting thoughts in your conscious mind, and it knows how to assemble the dresser even if you don't. This is because it's seen and analyzed the picture on the box and has compared it with other dressers which it had seen in the past.

2

You may have inadvertently looked at the instructions or even just part of the instructions, possibly for just a split-second. Your subconscious has a photographic memory and is quite capable of reading because you know how to read. You and your subconscious share the same memory.

So the knowledge of how to assemble the dresser was inside of you the entire time. Your subconscious was able to bring the thoughts to you because it knew where to retrieve them, and it did this because it wanted to help you.

It wanted to help you because you may have thought to yourself, "I hope I can figure out how to assemble this dresser." When you did that you were essentially communicating with your subconscious much the way it communicates with you. Your subconscious knew you wanted help and it wanted to help you.

That being said if you had thought to yourself, "I will follow the instructions and assemble the dresser myself," your subconscious would leave you to do just that.

If you were trying to do it by yourself and thought "I hope I can figure this out." then at that point your subconscious may start helping you.

If you simply already knew how to assemble the dresser because you had assembled other dressers in the past, you may not even think to yourself (communicate with your subconscious) and therefore it wouldn't help you. Your subconscious wants to help you only if you want its help. It is a part of you and will help you in anyway it can, whenever you need or want help.

To recap on things your subconscious is extremely good at solving problems. It has a hell of a memory. It will protect you, and it will help you whenever you need help. Your subconscious is a part of you.

Hypnotism is a way of putting your conscious into a waking-dream state, similar to sleepwalking; and allowing another person to verbally communicate directly with your subconscious. This is generally safe because your subconscious protects you and therefore won't do anything that you would ordinarily consider dangerous.

The subconscious is logical and will not get angry, sad, embarrassed, or overly emotional. Your subconscious is always positive and generally rejects negative thoughts, even when you're hypnotized.

For example let's say that you go and get hypnotized to help you quit smoking, and while you're hypnotized the hypnotist uses all positive phrases such as, "You will be happier and healthier after you stop smoking. You will feel more energetic, have more money, and your teeth will be cleaner. Etc." After having a positive hypnosis session like that, if you really want to quit smoking your subconscious will make it easy for you to do so.

This is because the hypnotist explained to your subconscious how quitting smoking will help you, or otherwise be good for you. Your subconscious only trusted that hypnotist because he/she never said anything negative throughout the entire session. On the other hand if the hypnotist had of used any negative wording, than your subconscious would not have trusted him and therefore would either ignore everything he said, or do the exact opposite.

An example of negative wording: "By quitting smoking you will be reducing your risk of developing chronic diseases such as lung cancer." This suggestion to the subconscious that you are currently at risk of developing a chronic disease is extremely negative because, whether or not there was actually any risk, your subconscious has been given an idea and that idea is "that you're currently at risk." Your subconscious will not hurt you so it will ignore the instruction, "develop a disease such as lung cancer."

Since the subconscious viewed this statement as negative it will ignore everything the hypnotist has said or will say as it doesn't trust him/her.

You may not personally find the above statement to be negative, but keep in mind that your subconscious looks at things from a more logical stand point. The subconscious is a complex and wonderful thing.

The subconscious sums up most situations instantly and generally plants a thought in one's mind when there's a problem. This is known as instant intuition or sudden understanding. For example; let's say that you were about to buy a used car and the first thought that pops into your mind is, "This car has been in an accident."

The person selling you the car swears it's never been in an accident of any kind, and the car to you appears to be clean and very nice.

It doesn't appear to you like the car has been in an accident, but somehow you just don't feel right about buying the car without more information, so you ask the owner who is selling the car "as is" if you can take the car for a pre-safety before making a decision to purchase it. The person selling the car says no, he claims there are other people interested in the car who may be coming to see it. So instead you write down the vehicle's identification number and leave.

When you get home you go onto the Internet (go online) and check the car's history on a website such as www.carfax.com. To your surprise you find out that the vehicle that you had been looking at has been in not one, not two, but three accidents, and you're quite relieved that you decided not to purchase that vehicle, without checking into things first.

Your subconscious sums things up instantly and is usually right. That's why most people often go with their first thought, or say things like, "Your first inclination is usually right." More on the subconscious in the next section.

Subconscious Communication

Have you ever wondered where your dreams come from? Were you ever amazed at how accurate a renowned psychic's statements actually were? Have you ever been in a situation where you were about to ask someone something, but amazingly they spoke about it before you could ask? You are about to read and fully understand exactly how and why these things happen/occur.

When you fall asleep many things in the body occur. The part of your conscious mind that sends signals to the rest of your body ceases to do so, as it goes into a state of rest; meaning that while asleep it would be difficult to move.

When in a deep sleep, it is nearly impossible to move as the part of your brain which your conscious uses to move parts of your body is almost completely at rest, and during your deepest states of sleep it is completely at rest and therefore any conscious movement of your body would be impossible.

This unmovable state is known as paradoxical sleep or REM sleep (REM = rapid eye movement).

During REM sleep people dream, however, other things are also occurring during the dream which will be discussed shortly.

When dreaming the subconscious becomes nearly 100% active, and the conscious is only about 3% to 5% active, and without the guidance of the subconscious.

During REM sleep, rather than the subconscious helping you to figure things out or make decisions, it is leaving you to do these tasks on your own; as it already has a task to do. That task is assembling or creating your dreams. How these dreams play out is generally left to the conscious mind, which again is only about 3% to 5% active.

It's as if you're subconscious mind has wrote a "choose your own adventure novel" for your conscious, and your conscious is reading it, and making choices while impaired (3% to 5% active), or similar.

Where do dreams come from?

As discussed earlier they are created by the subconscious, to keep your conscious occupied while your subconscious does other things. Your subconscious creates your dreams in a positive way using things it believes you will find familiar, interesting, challenging, or things that it knows that you always wanted to do, or try, but never have.

For example: During the day you may have seen a pregnant co-worker at work, seen a movie about the after-life. Not to mention you may have been worried about making it to work on time because you slept in a little and you may have thought to yourself, "I hope my old car gets me to work and doesn't brake down." Many other things may have happened during the day, but those things just mentioned were things your conscious mind was fixated on, thus they're the things your subconscious will likely use in your dreams. It may of course use things from other days, years, or even other dreams if you were fixated on (occupied in thought over) those things.

When you fixate on something you are communicating/sharing with your subconscious.

As an example of a possible dream made of the previously discussed conscious fixations/thoughts; you might dream of being in an old house with a pregnant lady who is or resembles your co-worker. You may try to leave for work, but then your car brakes down, so you return home (keep in mind it's like your impaired).

Once back at home the pregnant lady who resembles your co-worker may have called in a psychic to communicate with her dead mother, then things in the house may start floating around as you had seen in the movie which you watched earlier that day, and then at that point you would probably wake up.

If you remembered your dream you would probably figure out that most of the components of it came from things which you either have recently thought about deeply, or were fixated on.

However keep in mind that sometimes the subconscious will use things that you had thought about deeply (were fixated on) in the past, or even from other dreams; especially if, for example, you weren't deep in thought or fixated on enough material during the day to create a dream with when you went to sleep that night.

REM/paradoxical sleep is the most common type of sleep that the majority of people are aware of having been in from time to time, whether they can remember the dreams or not.

This brings up an additional question. If people are going to have dreams and not remember them, why have them at all?

Before I answer this question let's recap on what's happening during REM sleep.

During REM sleep the part of your conscious mind that allows you to consciously move your body is "off" as is 95% to 97% of the entire conscious mind. At this same point your subconscious mind is nearly 100% active.

It takes your subconscious about 2% to 3% of its resources to regulate and repair the body, and an additional 2% to 3% of its resources to create and maintain a dream for your conscious mind.

So to properly answer the question, "Why have dreams if you can't always remember them?" I'd have to say we have dreams to keep the conscious occupied, so that the subconscious is free to socialize and communicate with the subconsciouses of other people.

How is this possible you ask? Well as I mentioned earlier much of the subconscious is active/awake and free, during REM sleep, and the conscious no longer has access to the part of the brain which it uses when awake to control the movements of the body. So the subconscious is free/able to use that part of your brain for other purposes such as communication.

Think of the nerves in your body as ordinary household extension cords, as they are great at transferring electricity; however, ordinary household extension cords also make great antennas. Much the way in which we could use an extension cord as an antenna, you're subconscious is able to use your nerves as an antenna.

Your subconscious can transmit and receive UHF (ultra high frequency) signals directly to anyone else's subconscious, providing that person is physically located within about 200,000 meters (200 KM) from the other person.

Note: While a person is awake and fully conscious that persons subconscious can still communicate via UHF, however it is difficult for the subconscious to do this as that person's nerves are in constant use, as are the nerves in the other party. Aside from it being difficult, the range of communication should it occur, is much more limited (about 2000 Meters or 2 KM).

So to recap on things; when you enter REM sleep you can't move and your subconscious is using your body as an antenna to communicate with other people's subconsciouses, via UHF (ultra high frequency).

At this point you may be wondering what sort of things could one's subconscious possibly have to discuss with others. It is likely discussing ideas, creatively inventing things, as well as coming up with solutions to your and other peoples' problems, etc.

Keep in mind the subconscious looks at things quite differently than the conscious. The subconscious looks at things from a positive standpoint. Not to mention the subconscious has so many powerful resources, (memory, logic, analyzational skills, etc.) that it can some up a situation almost instantly.

That being said there is of course one major similarity between the conscious and the subconscious, which is, feelings of comfort and discomfort. For example, back in the day when running water just came out, but before toilets were invented, people may have experienced some discomfort when going to the washroom. Both their conscious and subconscious felt this discomfort.

Someone's subconscious somewhere invented the toilet, and then communicated that idea with as many other peoples' subconsciouses as possible, while the consciouses of these people were in REM sleep. Eventually many people began dreaming about toilets, which at that time didn't exist. This is one of the ways the subconscious communicates with the conscious.

Eventually some of the people who dreamt about toilets remembered their dreams, and eventually one person decided to build one. The invention of the toilet didn't just satisfy our discomfort, but also the discomfort of the subconscious.

Who's to say who really invented what? Can you really say for certain that Thomas Edison invented the light bulb, or that Karl Benz invented the first gasoline powered car in 1885? One thing I'll admit to is that I really can't be sure of who actually invented what.

I'm certain that a lot of our inventions were created by the subconscious. In addition to inventing things, the subconscious also discusses current and future events because it likes to figure things out and solve problems/resolve issues.

To recap on things a bit; the subconscious shares your comfort and discomfort. It invents and communicates with other peoples' subconsciouses while you sleep. It can communicate/demonstrate an idea to the conscious in the form of a dream, during normal REM sleep.

There are three other types of sleep/dream states, beyond REM, which I will touch on below:

Lucent dreams are a state of deep sleep in which the part of the mind that the conscious normally uses during REM sleep, is much more active than it usually would be.

Typically during a lucent dream the parts of the brain which are used during normal consciousness are about 55% active, plus or minus 10%.

Although your subconscious is providing and essentially creating your dreams, you are more or less conscious throughout the experience. You have a great deal of control over what happens during these dreams and thus what you end up experiencing.

For example: If you decided to go hang gliding during a lucent dream, you would experience exactly what hang gliding feels like, regardless of whether or not you've actually ever done any hang gliding. The subconscious is able to recreate this feeling because it has communicated with many other peoples' subconsciouses who have gone hang gliding.

In the unlikely event that it had never communicated with other peoples' subconsciouses who have gone hang gliding, then it would simply create the experience of hang gliding from scratch based on what it knows about the properties of flight and gravity; which is most likely as accurate as the actual experience.

Another example of lucent dreaming in which a person experiences something new, would be a virgin, who has dreamt about and experienced sex in a dream for the first time with someone whom he/she finds attractive.

In this sort of dream he/she will actually feel exactly as if they are having sex although that person never physically has, with much the same accuracy in the feeling as the person who chose to go hang gliding in his dream, but never had in real life.

During lucent dreams there is essentially a whole world out there for one to explore, which is extremely realistic, and without consequences, meaning; even if the person chose to rob a bank, murder someone, rape someone, or do anything considered illegal that person would get the thrill and excitement that he/she would have in real life, including a run in with the cops, but in the end would wake up and be perfectly fine as nothing ever actually physically occurred.

Lucent dreams can be either positive or negative such as a nightmare, but either way although they feel real, they are just dreams. Additionally, keep in mind, when a person is having a lucent dream that person is more or less conscious, thus that person knows he/she is dreaming.

This is obvious to that person although everything feels real, because many things will likely occur which are far from ordinary. Keep in mind even lucent dreams start out as ordinary dreams, and dreams are interesting places/experiences; anything can happen in a dream.

Less than 5% of the population (of Earth) experiences lucent dreams during their life time.

The next type of dream state which I would like to touch on is void state.

Void state is where a person was having a normal REM dream, but the parts of his mind which his conscious uses shutdown completely, thus the person's conscious is completely off (the person is completely unconscious) and therefore that person is not even dreaming. That being said the subconscious at that point is most active, and is communicating with other peoples' subconsciouses as previously discussed.

Anyone who does not have extreme brain damage will enter void state at least once per night, this is completely normal. That being said it would be extremely rare and out of the ordinary for a person to become conscious while in void state, however, it does happen. On average it happens to every second person once during their lifetime, meaning about half the population has or will consciously experience void sleep.

What is it like to be in void sleep, while conscious? Well to answer that, many things are happening when this occurs; primarily a person who becomes conscious while in void sleep will quickly notice that he/she is unable to move. That person will also usually be unable to open his/her eyes, not to mention that person's ears will be "off" so to speak.

All this is occurring because that person's subconscious is using the network of nerves in the body as an antenna for communication purposes.

So to recap on things a bit when a person is conscious and in void sleep he/she cannot move at all, and in fact probably can't even feel his/her arms, legs, head etc.

All that being said a person who is conscious while in void sleep may experience what are considered to be auditory and visual hallucinations. These hallucinations are mainly the result of the conscious mind picking up on the discussions that their subconscious is having with others.

The hallucinations may come in any form; from the person in conscious void sleep actually hearing the subconscious talking and having a conversation, to something that sounds like satellite radio transmissions.

As for the visual hallucinations, they generally are very limited. A person in this state who is having a visual hallucination may see a bit of white/colored light, movement of white/colored light, or may see a very black area moving amongst the darkness. That black area would appear to be blacker than dark.

These hallucinations are nothing to be afraid of, because again, these are only being caused by the conscious picking up on the subconscious communicating with other peoples' subconsciouses.

It is possible these hallucinations occur because the conscious tried to move its body while the subconscious was using these same nerves as an antenna for communication purposes.

The final state of sleep which I would like to mention is trance. Trance is much like the state your in when your having a lucid dream in the way that the part of the brain which the conscious uses is about 55% active, however it differs from other dream states in the way that a person in this state may be able to consciously move his/her body somewhat, and may be able to hear, feel and see etc.

What is trance? Trance is where the subconscious purposely communicates directly with the conscious. Just imagine being able to ask your subconscious anything, and getting an instant accurate response!

Trance is the rarest state of sleep; only about 1 in 10,000 people ever get to experience this fantastic state of mind. Many cures for diseases and major inventions may have been thought up during trance.

Other forms of subconscious communication:

Other than just communicating with the conscious through dreams, or by planting a thought in one's mind, the subconscious can also communicate physically. For example two people may be having a conscious, normal discussion about something. While they're talking their subconsciouses may communicate with each other via body language.

When our subconsciouses are communicating physically in this fashion it is usually quite subtle. For example minor repeated eye movements, facial expressions and other idiosyncrasies. In addition to communicating via body language the subconscious also can communicate by talking backwards, as you're talking forwards, sometimes this can cause people to slur their speech.

The subconscious usually won't communicate in these fashions (body language and reverse talk) unless it is trying to obtain information for the conscious. For example: If you thought you misplaced something at work, but were unable to find it, your subconscious may communicate with your co-workers' subconsciouses to find out what happened to it; only because it knows you didn't misplace it (because it was there), and it knows that whatever you thought you misplaced was important to you.

If the subconscious of the person in question finds out what happened to the missing item it will usually make several attempts to communicate with the conscious, usually just by planting a thought in the mind, but sometimes by other means as well.

Another example would be two people talking about something, and then all of a sudden they both mention something unrelated at the same time. Their subconsciouses were communicating with each other and both planted the same thoughts into the conscious at roughly the same time.

Sometimes there is even three-way communication, for example if a person found another person attractive, and wondered to himself whether or not the person that he found attractive was available; than during a conversation with that person both peoples' subconsciouses might freely discuss the situation and then his subconscious may plant a thought in his mind to let him know whether or not she's single, while her subconscious may plant a thought in her mind so that she may inadvertently mention it. (e.g. "Me and my boyfriend . . .").

Three-way communication (conscious to subconscious, subconscious to subconscious, subconscious to conscious) is quite common and happens everyday, although most people are unaware of it.

The most powerful example of three-way communication would be when a person believes that he/she is a psychic and has a reputation for being extremely accurate with his or her predictions.

The reason for the astounding accuracy of his/her predictions would be three-way communication (conscious to subconscious, subconscious to subconscious, subconscious to conscious).

The psychic asks his subconscious about something (by focusing on it, or thinking deeply about it), his subconscious then talks to the other subconsciouses of the people visiting him via physical communication (body language etc.); then the psychics subconscious plants thoughts into his conscious; in the form of either words, sounds, or images.

When the psychic consciously tells his guest what he believes is the truth, they are amazed, because it is.

From a conscious standpoint the psychics conscience has analyzed his guest consciences (via three-way communication), but there was nothing really spiritual or magical about it. However sometimes it can seem that way, because three-way subconscious communication is generally quite subtle, unless one is aware of it and watching for it (body language, tone of voice etc.) and even then it's still quite difficult to detect.

In summary the subconscious is a highly organized, highly intelligent, positive and helpful entity. Each person has their own subconscious. The subconscious can some up a situation instantly and can communicate by; planting thoughts in the conscious, physically (3-way), and electrically via UHF (ultrahigh frequency) signals during

REM sleep. Your subconscious is a part of you and will help you whenever you need or want help. The subconscious loves a challenge, and loves working out problems.

Whenever you are thinking hard about something, or occupied with something, you are in essence communicating with your subconscious, as it knows what you know. Your subconscious will keep thinking about whatever you were occupied with, long after you give up on it, and when it comes up with and answer or solution it will plant that thought in your mind. This is often thought of as instant intuition/sudden understanding.

For example: Let's say you forgot something and couldn't remember, you gave up on remembering it, and then later the thought just came to you. Or you were thinking about what color to paint your room, but then got distracted by a phone call, before you even finished the phone call you had already more or less figured out what color you wanted. This is because the subconscious continued to work on finding the forgotten memory, or continued thinking about what color your room would look best.

In conclusion subconscious communication is responsible for many of the things we would ordinarily take for granted, such as inventions; or things we couldn't otherwise explain such as psychics, ESP etc. In any case learning to communicate effectively with your subconscious is a useful and masterful skill, one that most people are truly oblivious to.

Thank you for reading my thoughts and theories on the subconscious, and on subconscious communication. I really hope you enjoyed them, or at the very least found them interesting.

In the next section of this book entitled "The Universal Equation" I will be releasing and fully explaining the universal equation. I'm very confident that this concept will be completely new to you, either way I'm sure you will find it both fascinating and easy to understand.

The Universal Equation

The universal equation is E=M (Energy = Matter).

Energy and matter are one and the same. For example gravity, light, electricity, magnetism, radiation, heat, momentum, kinetic, and potential are all forms of energy. All of these forms of energy involve matter in motion, even potential.

Electricity is the motion or flow of electrons, but consider that electrons are always moving even in a battery which is considered to have/hold chemical potential energy. Electrons are matter as is everything else tangible to us.

You might be under the impression that energy is mass in motion or "E=M*M", but this is truly not the case, because all matter is in motion to some degree. For example an ice cube is made up of water molecules in motion which are held in a variable fixed proximity by micro-gravity and anti-micro-gravity.

In the example of a compressed spring (potential energy) micro-gravity and anti-micro-gravity particles are moving so that the molecules that the spring is made of are constantly applying force to regain their previous positions. Even though the spring remains compressed, it is constantly releasing energy (free energy) as force to return to its uncompressed state, as well as having what is considered

18

potential energy, because if you were to release the spring it would expand.

Energy and matter are essentially the same thing, we just tend to think of energy more as something which we can use, rather than something that is all around us at all times.

No object can have a temperature of absolute zero, as will be discussed in just a few moments.

Energy & Matter. (How they tie together.)

As explained in the universal equation E=M, energy and matter are the same thing. One might call electricity or light energy, but electricity is simply the flow of electrons and light is just shattered electrons, both of which are physical particles in motion.

Momentum is referred to as energy, as is heat, and both are further examples of particles in motion.

All that being said, all matter is always in motion. Some scientists will argue this with the theory of "Absolute Zero" meaning a temperature so cold that all atomic motion stops.

As will be explained in the up coming section of this book entitled "The interactions between Gravity, Micro-gravity, and Ultra-micro-gravity" all things are kept together by gravitational forces. Also as I'm sure you have observed sometimes when certain objects get really cold, especially plastics, they become brittle.

This is because there is more available micro-gravity present in the plastic that is cold than there would be in plastic at room temperature; which causes the plastic molecules to have a shorter variable fixed proximity from each other, which effectively causes the plastic to become harder.

However, because there are still the exact same number of molecules, and the molecules have for the most part shorter variable fixed proximities from each other, some spots in the plastic will have microscopic holes/gaps and these holes can cause the plastic to become brittle.

If it were possible to freeze an object to absolute zero than everything the object was made of including the atoms and components of them would all come apart, because micro-gravity is extremely small particles in motion, and if they were not moving because of a lack of temperature (motion), then there would be nothing left to hold everything together so it would immediately fall apart. As it falls apart what you have is particles in motion (specifically micro-heat, and heat), and at that point it would immediately stop falling apart as heat, micro-gravity, and gravity are produced.

The point being that all particles of matter are always in motion. Energy is defined as particles in motion. In that way energy and matter are both the same thing, not to mention they are both comprised of the same material; and as we will cover in the later section of this book "Nuclear Reactions and Radiation" the energy released from nuclear fission is equal to the loss of mass. The loss of mass is effectively just the loss of matter. Energy and matter are precisely the same; therefore, the universal equation is E=M (Energy = Matter).

Gravity. (What it is. How it works!)

Have you ever wondered what gravity actually is, or how it really works?

What is gravity?

Both Newton and Einstein essentially theorized, to put it simply, that gravity is the attraction between two masses.

Gravitational force is currently calculated as "Gravitational Force = (G * M1 * M2) / D²"; meaning gravitational force is equal to gravitational constant (6.67 * 10E -8 dyne), multiplied by mass 1, multiplied by mass 2, divided by distance squared.

I do agree that the approximate force of gravity can be calculated in that fashion; however I disagree with what both Newton and Einstein state that gravity is.

Gravity is not the attraction between two objects! The objects push themselves towards each other.

Before explaining why this is, let me just state that I can explain exactly how this works. Please keep in mind that neither Newton nor Einstein can explain how gravity attracts objects or holds you on Earth. That's because it doesn't.

Most people will say you're held down by a gravitational field, but if asked how it holds you down the response generally given is, "No one really knows," which clearly does not answer the question.

Everything has some gravity, gravity is the release of particles about one/two billionth the size of an atom. Gravity pushes objects together. Now if you're smart you should be thinking to yourself, "Yeah right, How?"

Answer: To make this is simple as possible, picture two perfectly spherical objects. They could be a planet and a star, two molecules, two marbles, or two peas etc.

Now picture each of the two spherical objects simultaneously releasing hundreds of billions of tiny particles in all directions constantly.

Since the particles are being thrown away from the spherical objects in all directions; neither of them has any reason to move, but as the objects get closer to each other, some of the tiny particles being thrown off the first spherical object brake some of the tiny particles on the second spherical object before they can be thrown off, and some of the tiny particles on the second spherical object brake some of the tiny particles on the first spherical object before they can be thrown off as well.

This causes a reduced force between the objects (where less particles are being thrown off), so the two objects have more force pushing them together than in any other direction, regardless of orientation.

Unless another force or physical obstruction stops the two spherical objects from moving, than they will move closer together.

As the objects get closer together more of the tiny particles from the first object smash more of the tiny particles on the second object, before they can be thrown off towards the first object and vice versa.

This causes an even grater decrease in the force between the two objects, so they push themselves towards each other even harder. The closer they get to each other, the grater the affect of the gravity.

You can now imagine/understand how objects push themselves together, rather than being held together by some magical force, pending your belief in the tiny particles one/two billionth of the size of an atom.

Note: It will be clearly explained where these particles come from and what causes them to move in a later section of this book entitled "Dimensional Sciences."

So to recap on things, tiny particles are being released from all matter on a subatomic level, and these particles are being released in every direction. So when two objects get close enough together some of the particles being thrown off each of the two objects smash some of the tiny particles on each other before they are thrown off, causing a decreased force (decrease in outward force) between the two objects. This causes the objects to want to move closer to each other so to speak.

The closer they get towards one another the less force there is pushing them away from each other, and the more they want to move even closer together; until the point where they are almost touching, which is where the greatest amount of gravitational force occurs.

The two objects won't actually touch each other, they could under extreme circumstances such as an explosion but only for a split-second because if they were to touch then the particles being thrown off both objects will push them apart until they reach a distance where gravity will kick in again. That distance is about ten billionths of a micron. Note: The particles start to push the objects apart before they touch each other, this causes a fixed distance between them. The fixed distance will vary based on the mass and size of the particles in question (variable fixed proximity).

Every tiny particle of visible matter has its own gravity, gravity on one particle is not limited to just pushing objects in one direction, but rather in all directions.

Gravity causes our bodies to push/hold themselves together. Gravity pushes/holds all matter together. It pushes us down towards the Earth, and pushes/holds the Earth towards the Sun, while the Moon pushes itself towards the Earth.

When the Moon travels in-between the Earth and the Sun during an eclipse, the Moon wants to push itself both towards the Earth and the Sun as there is a reduced force in-between the Sun and the Moon as well as in-between the Moon and the Earth. This results in the two forces cancelling each other out, so that the Earth still wants to push itself towards the Sun against centrifugal force and continues to hold its orbit. This same sort of thing also occurs on a molecular level.

Gravity is also what holds the Sun together. Each exploding nuclear particle pushes itself back towards the Sun, unless the particle is made up of such a small amount of matter that it cannot produce enough gravitational force to push itself back towards the Sun's powerful nuclear explosions; which is the case with light, heat, and radiation.

Note: Additional information on light will be provided in a later section of this book entitled "Light & Vision (What it is. How we see it.)"

These particles have been accelerated to about 600 miles per second from the explosive force of the Sun; however, these particles continue to accelerate until they reach a peak speed of about 187,000 miles per second. The reason for this continued acceleration is that these particles do not contain enough matter, to have enough of their own gravity to hold them selves together. Thus they accelerate as fast as possible via dispersion.

Light, heat, and radiation are all low gravity emitting particles, and are constantly dispersing from the moment of their creation. The reason they travel at approximately 187,000 miles per second and not at an unlimited speed is due to dimensional limitations, which will be further covered in a later section of this book entitled "Dimensional Sciences."

To try and give you a better understanding of the intricate interactions between gravity and matter I will give a few more examples below:

Ex. 1) All the molecules in your body are pushing themselves together to form cells, every molecule in each cell is also pushing itself towards every molecule in every other cell causing your body to hold itself together, but while this is happening, every molecule in your body is also pushing itself towards the Earth.

Ex. 2) In drips of rain, the water molecules are pushing themselves together, but simultaneously pushing themselves towards the Earth.

Ex. 3) Think of 2 ocean liners or large ships in the ocean within close proximity of each other. The ocean water is pushing itself towards the Earth and towards the ocean liners to a certain extent. Every molecule of the ocean liners are pushing themselves together, while simultaneously they are pushing themselves towards the water and the Earth.

The Earth is so large and has so many gravity particles that it not only causes all the dense water molecules in the ocean to push themselves towards it, but many of the Earth's gravity particles escape through the water to cause the molecules in the ocean liner to push themselves towards the Earth. Note: That the gravity between the molecules of the ocean liner is stronger than the gravity between the Earth and these same molecules. This is solely due to the distance between the molecules.

While the gravity of the ocean liner is causing it to push itself towards the Earth, the gravity of the Earth and Sun is traveling through the ocean liner and ocean, to cause a decrease force between the Earth and the Sun, so that the Earth will want to push itself towards the Sun against the centrifugal force of its orbit, causing it to hold orbit.

The reason the Earth doesn't push itself right into the Sun is because of centrifugal force.

What is centrifugal force you ask?

Answer: All objects with mass always want to continue traveling in the same direction that they were headed in. There is a natural resistance to a change in direction based on mass. Note: This force is not related to gravity.

To give you a brief example of centrifugal force and its affect against gravity, let's say that you fill a bucket with water and hold it in your hand, then swing the bucket in circles without moving your legs. You will observe how the water stays in the bucket even when the bucket is upside-down over your head, and despite the fact that the water in the bucket wants to push itself towards the Earth. This is again because the water wants to travel in a straight line, and its resistance to deviate (due to mass) from the circular motion, causes it to stay in the bucket.

In conclusion gravity is a wonderfully complex interaction of subatomic particles causing every bit of matter to push itself towards every other bit. The amount of matter an object consist of, determines the amount of gravity particles emitted, and thus the overall approximate force of gravity.

Gravity particles also move through solid objects, for example a stack of weights at a gym. Gravity from the Earth causes each weight to push itself towards the Earth, and each weight has enough gravity to push themselves towards each other, but not to any appreciable extent. Like light and electricity, gravity particles also travel at about 187,000 miles per second. This high speed is caused by particle disbursement, and is limited by dimensional limitations. More on this in a later section of this book entitled "Dimensional Sciences."

Gravity particles are far smaller than light particles, thus their invisibility. Most people need to see something to believe it, but not everyone. I'm not saying you should blindly believe everything someone tells you. Please do doubt them, unless of course they can give you a plausible answer of how and why, one that you can understand.

In the next section of this book we will be thoroughly covering the topic of magnetism, including explanations of how and why magnets do what they do.

Thank you for taking the time to read my thoughts and theories on gravity; at the very least they should make you think.

__Magnetism__

What is magnetism?

From an early age in school we are taught that magnetism is the attraction of opposite magnetic poles, that like poles repel from each other, and that these magnetic fields are created from the alignment of invisible magnetic lines.

We are never taught how a magnet attracts certain metals, or why the poles on magnets will attract a repel each other, and if we ask a teacher these sorts of questions he/she will generally always answer by saying, no one really knows.

Although the world as of yet does not know how magnets work, people have been using them for hundreds of years in devices such as compasses, generators; and over the past hundred years in devices such as speakers, cars, telephones, computers, and medical devices such as an MRI (magnetic resonance imaging), etc.

You are about to read and finally receive an honest plausible answer as to what magnets actually are, and subsequently how they actually attract/repel each other, as well as why they attract various metals.

Since most everything you have been taught in school about magnets is/has been wrong; then what is a magnet really?

Answer: A magnet is an object packed with billions of billions of billions of magnetic particles (low gravity emitting particles). These particles have a natural variable limited release. Similarly the materials used in a battery contain electrons, which are also low gravity emitting particles.

All low gravity emitting particles such as electrons, particles of light, magnetic particles, particles of radiation etc. all move by dispersion and in just 1/50 trillionth of a second, they accelerate up to 187,000 miles per second. These particles cannot travel or disperse faster than 187,000 miles per second due to dimensional limitations, which will be discussed further in a later section of this book entitled "Dimensional Sciences."

We will recap and touch on exactly how a magnet works, but first I'd like to take a moment to better explain exactly what a low gravity emitting particle actually is, since low gravity emitting particles are key in my theories on magnetism.

What is a low gravity emitting particle?

Answer: Low gravity emitting particles are particles which are so small, that they produce virtually no standard gravity particles. For example light produces virtually no standard gravity particles, there are some, but so few that light is not restricted by gravity, not even the gravity of a star such as our Sun.

All materials in our dimension need gravity to hold them together. (More on that will be covered in the Dimensional Sciences section of this book.)

When something such as light does not have enough gravity to hold itself together it will disperse as fast as possible, which is again 187,000 miles per second due to dimensional limitations. On a side note gravity particles are not bound by the speed limit of our dimension, they are the one exception.

That being said, sometimes an object such as a magnet is made of multi-stage matter, meaning it contains/produces three or more sizes of gravity particles. In a magnet there are micro-gravity particles being emitted from each atom, thus holding it together, and normal gravity particles holding it on Earth. There are also ultra-micro-gravity particles causing some of the micro-gravity to group (grouped micro-gravity), and magnetic particles which are an additional rare size of gravity particle.

So to clarify on this, a magnet is being held together by micro-gravity, but it is also emitting several other sizes of gravity particles. Let's call these additional smaller gravity particles "ultra-micro-gravity" and the slightly larger ones magnetic particles. More information on gravity particles will be covered later in this section, as well as in additional sections of this book.

All objects emit some micro-gravity in addition to normal gravity. Both the amounts of micro-gravity and normal gravity emitted very with each object/substance.

Note: Ultra-micro-gravity particles are emitted by some micro-gravity particles but not all, ultra-micro-gravity causes micro-gravity to group (grouped micro-gravity).

Grouped micro-gravity is common in nature, but rarely ever seen in objects/substances with the exception of magnets and certain metals.

Let's compare a cube of plastic, a cube of aluminum, and a cube of cast-iron.

The cube of plastic has 70% as much matter as the cube of cast-iron, but pushes itself towards the Earth with less force than the cube of cast-iron because it is emitting less standard gravity particles. That being said there is also less micro-gravity holding the plastic cube together, thus it is easier to bend, brake, or melt plastic.

The aluminum block, on the other hand, is actually made up of 30% more matter than the cast-iron block, however it doesn't emit as many standard gravity particles as the cast-iron block, thus it is lighter as it is pushing itself towards the Earth with less force.

There is also somewhat less micro-gravity holding the aluminum block together, despite the fact it is made up of more matter than the cast-iron block, because of this the aluminum block has a lower melting point than the cast-iron block.

In essence micro-gravity holds an object together, while gravity pushes it towards the Earth.

So to recap on things and get back on track, as mentioned at the start of this section a magnet is an object that is packed with billions of billions of billions of magnetic particles (low gravity emitting particles). Most of these low gravity emitting particles are held in the magnet by micro-gravity, however some do escape creating a magnetic field; bringing us to the next question, "What is a magnetic field and how does it work?"

Explanation of magnetic fields, including how they work:

A magnet is an object which contains billions of billions of billions of magnetic particles (low gravity emitting particles, which are particles that are virtually unaffected by standard/normal gravity). The largest majority of these particles are held in place by micro-gravity; however some do escape creating a magnetic field.

When a magnet is placed near a metal object such as a quarter, the magnet pushes itself towards the quarter, and the quarter pushes itself towards the magnet. This is because the low gravity emitting particles (magnetic particles) being released from both the quarter and the magnet are acting as a type of gravity known as magnetism.

The scenario that I just mentioned brings up a couple more questions, such as:

1) What exactly causes the quarter to want to move towards the magnet, and vice versa?

2) If the quarter has low gravity emitting particles (magnetic particles), how come quarters don't attract each other?

To answer the questions of what's causing the magnet and the quarter to push themselves together, and why quarters don't push themselves together if they also give off low gravity emitting particles; let's take a look at what's happening with the magnet and then with the quarter.

The magnet is releasing billions of low gravity emitting particles (magnetic particles) in all directions, the quarter was releasing next to no magnetic particles (low gravity emitting particles), however the quarter does have a enough grouped micro-gravity that it can hold some of these escaping low gravity emitting particles (magnetic particles) from the magnet on its surface for at least a split-second; then some of the magnetic particles (low gravity emitting particles) begin to escape from the quarter's surface in all directions.

Both the quarter and the magnet want to move towards each other, because some of the magnetic particles from the magnet are smashing some of the magnetic particles on the quarter before they can be thrown off, and some of the magnetic particles from the quarter are smashing some of the magnetic particles on the magnet before they can be thrown off; thus causing a reduced force between the magnet and the quarter.

That being said if two quarters were in very close proximity or touching and then were placed near a magnet, then some of the temporarily emitted magnetic particles from the surface of the first quarter, which are escaping in all directions, may be caught by the micro-gravity of the second quarter for a fraction of a second (until it's surface can no longer contain all of the magnetic particles being thrown towards it), at which point its surface begins releasing magnetic particles in all directions.

Once this happens then some of the temporary magnetic particles from the first quarter will smash some of the temporary magnetic particles on the second quarter before they can be thrown off, and some of the temporary magnetic particles on the second quarter will smash some of the temporary magnetic particles on the first quarter before they can be thrown off. Thus causing a reduced force in-between the two quarters, which in turn causes them to move towards each other, until the source of the magnetic particles is removed.

Depending on the strength of the magnet (how many magnetic particles are escaping from the magnet) it may be possible for more than two quarters to share in the temporary magnetic field as previously described.

That being said to answer the question as to why quarters are not magnetic, even though they have enough grouped micro-gravity to hold magnetic particles; I'd have to say it is because they simply have never been exposed to powerful magnetic fields, or in other words they have never been magnetized.

Two ways to magnetize an object:

One is by exposing it to a powerful magnetic field for a long period of time. This works because some of the magnetic particles that the object is being exposed to are being absorbed by it, to the point that it has more than just a surface charge of magnetic particles, and therefore has become magnetic itself since it will be throwing off some of its own magnetic particles even when it is not exposed to a separate magnetic field.

The second way to magnetize an object is by rubbing a magnet over it in one direction, this will again cause magnetic particles to transfer from the magnet to the object being magnetized. This is because some of the magnetic particles are getting caught in the object being magnetized by its micro-gravity.

Let's recap on things and further explore what a magnet is. A magnet is an object that contains billions of magnetic particles, which are held in place by micro-gravity in what is known as the magnet's south pole.

The micro-gravity in the north pole of the magnet is grouped together by ultra-micro-gravity. Ultra-micro-gravity works like gravity but on a far smaller scale, and is simply the release of tiny particles from the material that the micro-gravity in question is comprised of. Ultra-micro-gravity acts as gravity for the micro-gravity and causes it to group.

Note: Not all micro-gravity emits ultra-micro-gravity; consequently few objects contain grouped micro-gravity with magnets being one of the exceptions. The ultra-micro-gravity affects only the micro-gravity and not the magnetic particles (low gravity emitting particles) of a magnet's south pole. This is the reason that there are two distinct polls.

So the reason there are two distinct polls is effectively because of grouped micro-gravity.

Similar polls repel because the escaping grouped micro-gravity in the north pole of one magnet is virtually unbreakable by other identical escaping grouped micro-gravity in another magnet's north pole due to the amount of matter it is composed of. So in other words the north poles' of magnets repel from one another.

They repel from each other because neither magnet's north magnetic field (escaping grouped micro-gravity) will brake to any appreciable extent, thus there is essentially a pressure build up that occurs between the north polls virtually speaking, this is not air pressure, but simply a non-tangible pressure to us. The pressures are caused by an enormous amount of escaping grouped micro-gravity particles bouncing off each other and escaping from a limited space. It is the limited space which causes the pressure to occur.

Similarly the magnetic particles of a magnet's south pole are too strong to brake to any appreciable extent when hit by identical magnetic particles from another magnet's south pole.

That is the reason that similar magnetic polls repel.

Opposite polls attract because the escaping grouped micro-gravity (north pole) brakes down the escaping magnetic particles (south pole), and the broken magnetic particles of the south pole brake down the escaping grouped micro-gravity of the north pole, causing a reduced force in-between the magnets which causes them to push themselves towards each other.

Why are only certain metals affected by magnets?

Answer: Steel like in a quarter has a high amount of grouped micro-gravity, whereas copper like in a penny does not. This is the reason only certain metals are affected by magnets.

Why does either pole of a magnet attract certain metals?

Answer: Since escaping magnetic particles of a magnet brake down the escaping grouped micro-gravity of certain metals, e.g. iron, certain metals are attracted to a magnet's south pole.

Additionally because the grouped micro-gravity of certain metals temporarily holds and then very rapidly releases magnetic particles, either pole of a magnet will attract certain metals.

While grouped micro-gravity holds magnetic particles, it can't hold them for very long because magnetic particles brake down grouped micro-gravity, thus the reason it can only hold them for a split second.

In a non-magnetized quarter the grouped micro-gravity is even throughout the quarter, and not escaping to any appreciable extent, unlike in a permanent magnet where the grouped micro-gravity is effectively off to one side (opposite of the magnetic particles).

If you temporarily expose a quarter to a magnet, both the magnetic particles and grouped micro-gravity are escaping evenly from every part of the quarter; as such, although the quarter is still magnetic, it has no distinct poles.

That being said there are less of both magnetic particles and grouped micro-gravity escaping from the non-magnetized quarter which is being exposed to the magnet, since some of the grouped micro-gravity is destroyed by the magnetic particles entering and temporarily being held in the quarter.

Similarly some of the magnetic particles in the quarter are destroyed by the grouped micro-gravity that is temporarily holding them; thus there is some magnetic loss in non-magnetized objects. For example, if you attach a quarter to a magnet, and then attach another quarter to that quarter, then another etc.; with each additional quarter there is more and more magnetic loss, which is the reason that a magnet will only hold so many.

Note: Put enough magnetic particles into a quarter (magnetize it) and the grouped micro-gravity will move off to one side, the magnetic particles will then occupy the other side and what you have is called a permanent magnet; although the magnetic field it releases will become weaker and weaker over time.

To clarify on things a bit, a magnet is made of a material which contains a high amount of grouped micro-gravity, and grouped magnetic particles. Metals which are attracted to magnets contain high amounts of grouped micro-gravity, thus it is possible to magnetize them (charge them with magnetic particles) as mentioned earlier.

Now that you have a better understanding of how magnets work, and why a magnet does what it does, let's explore how heat and impact affect magnets.

Why do high temperatures affect a magnet's strength?

When a magnet's temperature rises, mostly all grouped micro-gravity/magnetic particles are expelled/released, thus it is no longer magnetic. This happens because the molecules the magnet is made of begin to travel or vibrate rapidly, and the molecular sized spaces between the molecules inside the magnet that contain the largest amounts of grouped micro-gravity/magnetic particles are virtually eliminated due to this vibration. (Either the grouped micro-gravity/magnetic particles leave or are destroyed, as there is not enough space for them due to the vibration of the molecules.)

Note: The ungrouped micro-gravity that holds the molecules in a variable fixed proximity of each other lessens with heat, but otherwise remains intact. Heat causes a loss of grouped micro-gravity, and without grouped micro-gravity there is no magnet.

The ultra-micro-gravity being emitted by the micro-gravity will return to its previous state after the magnet has cooled, meaning that the magnet will still have the same amount of grouped micro-gravity, although not concentrated off to one side of it to nearly the same extent as before. Therefore the magnet may not have as many magnetic particles nor as much grouped micro-gravity being emitted from it, and thus less magnetic force. This is because the grouped micro-gravity and magnetic particles scattered within the magnet are destroying each other and therefore cannot be emitted.

Why does severe impact affect a magnet's strength?

When a magnet experiences a severe impact it loses strength for a similar reason. Grouped micro-gravity and magnetic particles have mass. Let's say you were to drop a magnet on a concrete floor, although it physically stops moving when it hits the ground, some or most of the grouped micro-gravity and magnetic particles continue traveling and essentially leave the magnet.

Another, yet more complicated way that a magnet can loose its strength during a severe impact is as follows. Let's say you strike a magnet with a hammer. The magnet as a whole accelerates and decelerates rapidly. All the atoms the magnet consists of move together.

This acceleration/deceleration is not enough to cause the grouped micro-gravity and magnetic particles to leave the magnet to any appreciable extent; however, it may lead to a chain reaction which would cause loss of many of them. This phenomenon is known as a grouped micro-gravity explosion, and this explosion includes/affects magnetic particles as well, since without grouped micro-gravity a loss of magnetic particles will occur.

Below you will find a brief explanation of what is happening during a grouped micro-gravity explosion, additionally this subject will be covered thoroughly in a later section of this book entitled "The interactions between Gravity, Micro-gravity and Ultra-micro-gravity."

What is a grouped micro-gravity/magnetic particle explosion?

Answer: As mentioned a moment ago when you strike a magnet with a hard object such as a hammer, the magnet rapidly accelerates/decelerates. (All of the atoms the magnet consists of rapidly accelerate/decelerate together.) The atoms are not vibrating amongst each other, that would be a temperature increase, but rather they are moving together synchronously.

Note: It is the micro-gravity that's being emitted by the magnetic particles that causes them to stay within the magnet (magnetic south pole), and it is the ultra-micro-gravity emitted by the grouped micro-gravity which causes it to group (magnetic north pole).

During impact, a rapid synchronous atomic acceleration/deceleration occurs. Also referred to as kinetic energy.

When the atoms are accelerating and decelerating synchronously and rapidly, so are the magnetic particles and grouped micro-gravity. If (highly likely) a magnetic particle or particle of grouped micro-gravity (1 out of 23547 billion) splits even a single other magnetic particle or particle of grouped micro-gravity, then that magnetic particle would no longer be held in place by micro-gravity as it would be smaller and its shape would be uneven. For example a half or quarter sphere.

Since the magnetic particle fragments are not being held by micro-gravity they would accelerate via dispersion to speeds up to 187,000 miles per second (dimensional speed limitation), which will be discussed in the "Dimensional Sciences" section of this book. Those fragments, because of there speed they would have enough force to brake other magnetic particles and even grouped micro-gravity. (Speed * mass = force).

The new grouped micro-gravity/magnetic particle fragments would then accelerate and go on to brake even more grouped micro-gravity/magnetic particles into fragments, those new fragments would do the same thing, and so on, and so on, until there are not enough magnetic particles or grouped micro-gravity particles to keep this reaction (explosion) going.

Once the reaction stops there would be so few remaining magnetic particles/grouped micro-gravity, that the magnetic force from the magnet would be so much weaker that most people would describe it as non existent.

Although I just talked about a micro-gravity explosion I was only referring to the grouped micro-gravity residing within the magnet, and not the non-grouped micro-gravity that holds various atoms together to form molecules.

Note: A non-grouped micro-gravity explosion is possible, amazing, dangerous and has been caught on film, and will be discussed in a later section of this book entitled "The interactions between Gravity, Micro-gravity and Ultra-micro-gravity."

A magnet that has lost its magnetic charge (magnetism) can be re-magnetized. This is because the micro-gravity that is being produced by the material the atoms are made of produces its own form of gravity known as ultra-micro-gravity, which causes the micro-gravity to group. Effectively the grouped micro-gravity quickly becomes replenished.

Again ultra-micro-gravity only affects the grouped micro-gravity that is producing it. No other materials are affected. Ultra-micro-gravity is what causes excess micro-gravity to group together, rather than leaving the vicinity of the atoms which are producing it. Atoms produce micro-gravity, which can be thought of as a byproduct of their existence.

Since this object has grouped micro-gravity it can be magnetized/re-magnetized (charged with magnetic particles).

All (even non-metal) objects are held together by micro-gravity, but very few objects produce grouped micro-gravity (micro-gravity which produces ultra-micro-gravity and thus groups). Thus the reason only certain materials can be magnetized (mostly metals).

Now that you have a better understanding of magnetism, you should understand that truly only the south pole of a magnet contains magnetic particles, and that magnetic particles are in essence just low gravity emitting particles, meaning that standard gravity has virtually no affect on them. (The north pole contains grouped micro-gravity).

Both the magnetic particles of the south pole, and the grouped micro-gravity of the north pole want to stay within the magnet, but they don't want to be anywhere near each other; thus the reason there are two distinct poles.

The space occupied by the grouped micro-gravity of the north pole; does not contain very many magnetic particles because it is jam-packed with grouped micro-gravity and the proximity of the micro-gravity that has grouped is so close that there is no/limited space for magnetic particles to be present; although micro-gravity attracts magnetic particles it attracts grouped micro-gravity more/harder. By attract I am speaking metaphorically of course. The magnetic particles within the magnet than occupy the remaining physical space "south pole."

You should also now be aware of three new concepts:
1) Micro-gravity, 2) Grouped micro-gravity, 3) Ultra-micro-gravity. All three will be further discussed in later sections of this book.

Thank you for taking the time to explore my theories on magnetism. I hope you enjoyed them, or at least found them different, interesting, or educational.

Electricity and the generation thereof

What is electricity?

Most people would say that electricity is the energy that powers devices in their home, such as a TV, radio, refrigerator, lights, dryer etc.

True electricity does all those wonderful things for us and more, but what is it really?

Most scientists, electricians etc. would say/state that it is the flow of electrons from positive to negative.

In almost all cases electricity is the flow of electrons from positive to negative, but not in all cases, and that explanation still doesn't really say what electricity is.

You are about to read and have the opportunity to truly understand exactly what electricity really is, how it is generated, how it works, and why it does what it does.

First off let's take a look at the many ways electricity is generated. There are nuclear, water, and coal power generating plants; as well as wind and solar farms. All of these, with the exception of the solar farm, have one thing in common; they all use generators to generate electricity.

A generator is a device with a rotating center that generates electricity via magnets, and which is cooled by air. It is very similar to the alternator found in a car, only much larger.

An alternator generates alternating current such as what you would find in standard wall plugs/outlets, where the positive and negative flow of electricity are alternating, the voltage is rectified into DC "direct-current" by a rectifier, then regulated to 12 Volts by a regulator.

So basically all electricity that we use today is generated by running magnets over a coil of wires.

Why does running magnets over a coil of wires generate electricity?

Before we examine the answer to this question, I would like to first touch on one simple point. "E=M" & "M=E" which means that "energy = matter" and "matter = energy." Energy and matter are one in the same.

Einstein stated that $E=MC^2$, meaning that mass at rest contained energy based on its mass, which is one way of looking at things, although not entirely accurate.

Newton speculated and theorized that a mass at rest had no kinetic energy, and relatively small amounts of chemical or thermal energy. Newton's theory is at least somewhat right, although it leaves much out. It pretty much leaves you with more questions than answers.

Einstein stated that a photon of light has no mass. This is not entirely correct, it has some, and I can justly say this based on the following examples:

1) Light is drawn into black holes (high gravity areas).

2) Light can act as a particle or as a wave.

3) The particles of light have mass, because all matter has mass. Energy and matter are the same thing (E=M). If something didn't have mass it wouldn't exist.

Both Newton's and Einstein's theories have some truth to them, but the absolute truth is that all matter has mass, and matter and energy are one in the same; thus my equations E=M and M=E (energy is matter and vice-versa). They are simple but true.

Similarly, practically everyone on Earth believes that electricity (electrons) always flows from positive to negative, however, that is a misconception and will be proven wrong using existing evidence in this section of my book.

More on the relationship between energy and matter will be covered in two later sections of this book entitled "Dimensional Sciences" and "Nuclear reactions and Radiation."

All that being said, let's get back to the original question. How does running magnets over a coil of wires produce electricity?

Answer:

When a magnet passes over a piece of wire the escaping micro-gravity from the magnet frees excess electrons from the atoms the wire is comprised of and also draws in free floating electrons from anything else in range of the magnetic field.

The metal wire has a surface charge of micro-gravity and the magnetic particles are extending out past the wire. As the magnet moves so does the stream of micro-gravity thus accelerating the electrons in the wire in the direction that the magnet is traveling in. Most electrons will stay within the field of micro-gravity and thus travel through the wire. (Electrons move from the greatest source of electrons to the greatest source of micro-gravity.)

Additionally some of the magnetic particles escaping from the magnet are getting caught in the micro-gravity of the wire and are compressed into electrons.

The flow of electrons is electricity, and the electrons flow because of micro-gravity.

To recap on things a bit, electricity is the flow of electrons (usually from positive to negative), and electricity (the flow of electrons) is generated through the use of magnet's in hydro generators.

As I just mentioned electricity is the flow of electrons usually from positive to negative, but not always.

Let's examine the term "from positive to negative", and then I'll explain how this rule is generally correct, however, is also a major misconception.

Electricity flowing from positive to negative means that electrons are flowing from an area with a surplus of electrons to ground.

That definition is simple and is generally correct, but as I said before, there are major misconceptions about it which I will explain below.

The truth is that electricity (electrons) flows from the greatest surplus of electrons to the greatest source of micro-gravity, using the easiest path possible (a conductor). A conductor could be a piece of wire, or an existing flow of electrons such as a radio wave/signal.

Generally scientists agree that electricity flows from positive to negative (ground).

To prove them wrong, and subsequently prove my theory that electrons flow from the greatest surplus of electrons to the greatest source of micro-gravity using the easiest path possible to be correct, let's take a look at lightning.

What is lightning and how does it brake the existing rules of electricity?

Lightning is a massive amount of electricity which sometimes flows from the sky (positive) to the ground (negative). Ground is considered negative by all electricians, meaning they believe it has a lesser or negative amount of electrons than positive.

Anyway, as I was saying lightning sometimes flows from the clouds to the ground (positive to negative); however, it is a fact that most of the time lightning flows from the ground to the sky (negative to positive). This fact can be checked online, or in modern books containing information on lightning. Note: To the average person looking at lightning, the streak appears instantaneously.

So why does lightning usually flow from the ground to the sky? (Negative to positive?)

Answer:

Lightning is electricity (movement or flow of electrons) and as I mentioned earlier it is a major misconception that electricity flows from positive to negative, as it does not. Electricity flows from the greatest surplus of electrons to the greatest source of micro-gravity using the easiest path available.

So to properly answer the question of why lightning usually flows from negative (the ground) to positive (the sky), I'd have to say it's because the ground usually contains a greater amount of electrons than the clouds in the sky, which happen to have massive amounts of micro-gravity.

The micro-gravity in the clouds causes the electrons in the ground to flow up towards it via a conductor such as a radio signal or moisture that is in the air during a heavy rain.

Due to all the micro-gravity in the clouds they also contain a lot of electrons, and sometimes if there is a path/conductor to a greater source of micro-gravity on the ground, the lightning will shoot down from the clouds, rather than shoot up from the ground; so sometimes the electricity that lightning consist of will flow from positive to negative, however usually it flows from negative to positive.

Let's take another look at electricity in motion and the complicated interactions thereof which most people are oblivious to, by examining a battery, any battery.

A battery is an object that is said to hold/store electricity (the movement of electrons) in the form of chemical energy.

In a simple battery we have two plates made of different/dissimilar electrically conductive materials. The two posts that we use to get electricity from the battery are either part of the plates, or attached to the plates. In addition to the two plates there is an acid. An acid is a material with a very low micro-gravity and or many electrons.

As mentioned a moment ago the plates in a battery are made of two dissimilar electrically conductive materials, and thus each plate has a different amount of, and slightly different size of micro-gravity. The acid itself also has a different amount of micro-gravity. Its free micro-gravity is based on the number of stored electrons.

When we charge a battery (add electrons), the electrons travel through the plates and are pushed into the acid. The micro-gravity in the acid holds these extra electrons; so there is less available micro-gravity to hold the material the acid is made of together, thus an inter-mixing occurs. Let's call this inter-mixing "micro-heat".

The micro-heat (rapid vibration & inter-mixing of the matter which the acid is composed of) burns away some of the material that the plates are made of. As it does this the acid gains this material and its micro-gravity, thus increasing the amount of micro-gravity in the acid, and thus lowering the amount of micro-heat/acidity.

If we stop charging the battery (stop adding electrons) the reaction will become stable, meaning that the micro-heat in the acid will burn away enough matter from the plates until the point that the acid has enough micro-gravity that it no longer has enough micro-heat to burn additional matter from the plates. By stable I mean the reaction stops.

On the other hand if we continue charging the battery (adding electrons) the reaction in the battery will continue to occur, as even though the micro-heat of the acid is burning away matter from the plates, adding more micro-gravity to the acid, and thus lowering the amount of micro-heat; the electrons being added to the acid are continuing to be held in the acids increased micro-gravity, thus lowering the amount of micro-gravity available to hold the matter the acid is made of together, thus increasing the micro-heat, and thus allowing the reaction to continue as long as the battery is being charged (as long as electrons are being pushed into the battery).

Let's recap on things a bit. When a battery is charged, the acid within it becomes more acidic (it has more micro-heat); meaning that the additional charge of electrons are consuming the available free micro-gravity within the acid.

The acid when highly charged with electrons will not freeze as easily because it has a high amount of non-conductive energy that I call micro-heat. The micro-heat exist because some of the micro-gravity which was previously holding the matter the acid is made of together is now holding the electrons. The matter the acid molecules themselves are made of is inter-mixing rapidly due to dispersion. This inter-mixing action is what I am referring to when I use the term micro-heat.

The micro-heat will cause some of the matter the plates in the battery are made of to mix with the acid and subsequently with each other. Note that the plates are made of slightly different materials/matter and thus they each have a different amount and slightly different size of micro-gravity.

The micro-gravity of the plates in the battery is causing the original material the plates were made of, which has been dispersed into the acid, to push itself back towards the plates; however it is being prevented from doing so by the micro-heat (due to the surplus of electrons).

When the battery is fully charged (has a surplus/saturation of electrons held in micro-gravity) micro-gravity still has a hold on the actual matter the plates and acid molecules are made of, but this hold is weaker since there is less micro-gravity available because it is being used to hold the excess electrons.

What pushes the electrons out of the battery during normal use? & What causes the electrons to flow in one direction over the other?

During normal use matter is traveling from the left plate to the right plate of the battery and vice versa, also plate matter inter-mixed within the acid is traveling back to its respective plates.

As soon is there is any electrical conductivity (use of a battery), some electrons flow from both the positive and negative post to the object/device being powered up. This is because electrons flow from the greatest source of electrons to the greatest source of micro-gravity using the easiest path available.

As soon as some electrons have left the battery there is more available free micro-gravity for the acid to reconstruct itself into a less acidic form (less micro-heat) and subsequently due to that fact, the particles of matter the plates are made of are able to push themselves out of the acid and return to their respective plates.

Electricity (the flow of electrons) appears to be flowing in one direction from positive to negative, but this is not the case. The matter in the positive plate has a greater size and greater mass than the matter in the negative plate. Electrons are flowing out of the battery from both plates at once, however more are flowing from the positive plate thus creating an indirectly-observable flow of electrons.

Keep in mind that electricity flows from the greatest source of electrons to the greatest source of micro-gravity using the easiest path available. So the terms positive and negative on a battery are truly only virtual.

Currently most people simply believe that electricity flows from positive to negative, however as mentioned above this is truly not the case at all. I submit to you the reader that my theory on how batteries work is correct as it explains how?, why?, and why not?

Static Electricity

Static electricity is nothing more than an accumulation of micro-gravity charged with electrons. Electrons release some of their own micro-gravity, thus the reason they push themselves towards other micro-gravity.

Magnetic particles and electrons both emit some micro-gravity. Sometimes micro-gravity emits ultra-micro-gravity thus causing it to group, as is the case with a magnet or static electricity.

The micro-gravity and electrons in static electricity are pushing themselves together because the micro-gravity is emitting ultra-micro-gravity. So what you have is micro-gravity emitting ultra-micro-gravity causing the micro-gravity to push itself together and group, as well as electrons which are also emitting micro-gravity causing them to push themselves towards the grouped micro-gravity.

As mentioned in an earlier section of this book entitled "Gravity. (What it is. How it works!)"; gravity is the release of particles from matter causing objects to push themselves together, micro-gravity works the same way but on a smaller scale. Ultra-micro-gravity is even smaller than micro-gravity and is comprised of particles which are disbursed from some forms of it.

More on micro-gravity will be covered in the next section of this book entitled "The interactions between Gravity, Micro-gravity, and

Ultra-micro-gravity." These interactions/phenomenon's are rarely talked about as they are not well known.

To conclude this section I'd like to say that electricity is simply the movement of electrons due to micro-gravity.

Thank you for reading my analysis of electricity and the generation thereof; I hope it opened your eyes and made you think.

The interactions between Gravity, Micro-gravity, and Ultra-micro-gravity.

Just what sort of interactions are there between gravity, micro-gravity, and ultra-micro-gravity?

This isn't an easy question to answer, however I will try to give an explanation after some examples, but first let's examine what micro-gravity is/does.

Micro-gravity works just like gravity. For more information on how gravity works please refer to the earlier section of this book entitled "Gravity. (What it is. How it works!)"

Micro-gravity particles are much smaller than gravity particles, and simultaneously can/do interact with the same molecules as gravity. A good example of this is a drip of rain falling from a cloud. Micro-gravity is causing all the water molecules in the drip of rain to push themselves together, while simultaneously gravity is causing the water molecules to push themselves towards the Earth.

There is one major difference between gravity and micro-gravity; sometimes micro-gravity emits ultra-micro-gravity, but not always, whereas gravity doesn't emit anything.

The micro-gravity that does emit ultra-micro-gravity will group together as it does in the case of a magnet, however keep in mind because micro-gravity can group it does not necessarily need to be emitted from an object; thus grouped micro-gravity could be anywhere, even in space, even in your bedroom above your bed.

You can't feel or see grouped micro-gravity, however it can be detected in other ways which I will try and demonstrate below.

As mentioned in an earlier section of this book entitled "Electricity and the generation thereof" we know that some micro-gravity is emitted from electrons, thus there are high levels of micro-gravity in the sky (usually) or even on the ground which are responsible for lightning.

Electrons by default can group together in an object which emits a high enough micro-gravity, such as the plates of a battery. If the electrons have a path to travel (a conductor) they will travel towards the source of the micro-gravity, and because of the nature of a battery (It consists of plates made of two different materials), there will be a grater observable flow of electrons from one of the battery's plates. Which ever plate has the grater flow of electrons is considered positive.

A blanket like most other things doesn't emit enough micro-gravity to draw electrons/hold electricity; however, sometimes when a blanket is in a dryer it may pick up significant amounts of grouped micro-gravity (micro-gravity which is emitting ultra-micro-gravity, causing it to group).

When free-floating electrons enter into the grouped micro-gravity some of them temporarily get caught. The cloth blanket itself may try and push itself towards itself or other cloth objects, as the cloths naturally emitted insignificant amount of micro-gravity is interacting with the grouped micro-gravity.

If you were to disturb the cloth blanket, for example; pull a sock from it, or unfold a part of it which was stuck to itself, you would otherwise be lessening the amount of grouped micro-gravity in a particular location to the point that some of the electrons would be free to escape and again become free-floating.

Also the Earth itself has a tremendous amount of micro-gravity, and if your body was standing in a location that was conducting enough of it sometimes rather than the electrons becoming free-floating, they would travel from the blanket or other object through you and into the floor in your house and finally to the ground of the Earth. You might feel a little shock from that, and usually you would be able to see a small spark.

What I have been explaining is known as static electricity; however, it is the grouped micro-gravity which is what makes static electricity possible.

An interesting example of micro-gravity is water. Your body emits enough micro-gravity to hold a drip of water to your finger, against the force of gravity which causes the water to want to push itself towards the Earth.

A much more complex example of the interactions between micro-gravity, gravity, and heat will be discussed in a moment; however, to make things more understandable, let me first explain the difference between heat and temperature. Most people think of heat and temperature as the same thing; however, they are entirely different.

Temperature is the vibration of molecules, whereas heat is a particle. Just like gravity, heat cannot be seen, but can be otherwise observed.

Both heat and temperature can be felt. If you touch something that is hot (something with a high temperature/rapidly vibrating molecules) you will feel the temperature increase in your hand. This is because some of the rapidly vibrating molecules in the object you touched caused some of the molecules in your hand to vibrate. This is called temperature transfer, and it is very different from heat.

To give you another example of temperature and an example of heat; let's say that you had a stove with an oven that had a window.

Let's say the oven was cold and you turned it on and waited five minutes. Then let's say after the five minutes you touch the glass window of the oven. When you do this you will observe that the glass window of the oven is still cold to the touch, meaning that the molecules in the glass window are actually still vibrating slower than the molecules in your hand, and when they connect/impact with each other the molecules in your hand slow down slightly, as they slightly speed up the molecules in the glass window of the oven. The feeling you get in your hand when you touch this glass could be described as cold, and was caused by temperature transfer.

With this same preheated oven that has a glass window which you know is cold to the touch, because you're touching it; when you move your hand back a few inches, then you feel heat. It would feel hotter then room temperature, although the glass is cold to the touch. This is because heat is a particle which is traveling through the glass and impacting on the molecules of your hand, thus causing them to accelerate/vibrate faster.

Heat is an obscure escaping particle which impacts molecules at 187,000 miles per second and thus causes them to vibrate. These molecules vibrate instead of travel because they are held in place by micro-gravity.

Note: Heat also causes a small, temporary decrease in the amount of active micro-gravity. This contributes to the rate an object expands when heated.

The temperature (vibration of molecules, due to heat) causes the object to expand, remaining/new heat particles cause some of the active micro-gravity to become ineffective either because it becomes broken, stops, or leaves the object altogether.

You can feel heat; heat causes temperature, and you can feel temperature. Both feel nearly the same. Both feelings are caused by the acceleration of molecules in your body. Heat feels slightly different than temperature because of the depth to which it accelerates molecules.

In other words heat feels slightly different than temperature because heat may for example cause several inches of molecules in your body to vibrate a little, whereas temperature transfer causes the molecules in your body to vibrate only on the surface. If you are exposed to a high temperature for a long enough period of time there may become more depth to the vibration of molecules in your body, but it's not instant like with heat.

Heat causes molecules to vibrate even if they do not hit each other. Heat can travel deep into an object and cause it to gain temperature more evenly than in the case of temperature transfer where one molecule hits another molecule and causes it to move.

Basically heat causes the vibration of molecules; temperature is the vibration of molecules, and temperature transfer is when one molecule hits another and causes it to vibrate. When I say "hit" I don't mean that they actually hit each other, however they get close to each other briefly and then push themselves away from each other and towards the molecule furthest from them due to anti-micro-gravity and micro-gravity respectively.

Now that I have explained heat, temperature, and temperature transfer; let's get back to the original scenario that shows more complex interactions between heat, gravity, and micro-gravity via the example below.

Clouds:

Have you ever wondered why clouds form, and how they stay in the sky against the force of gravity?

The reason clouds stay in the sky is because a single molecule of water weighs less than a single molecule of air.

Note: When billions of water molecules pack together in the same area, a drip of water is formed; and a drip of water weighs more than the equivalent amount of air which would otherwise be occupying the space, thus gravity is able to cause the drip of water to push itself towards the ground.

Heat from the Sun is preventing drips of water from forming as it is vibrating/accelerating the water molecules of the cloud.

Micro-gravity is causing the water molecules to try and push themselves together to form drips of water, but heat is accelerating/ vibrating them to the point that they cannot, thus the reason we have clouds.

Clouds have a high amount of micro-gravity, so high in fact, that when two or more clouds get close enough to each other they will usually push themselves together to form a single cloud regardless of the direction of the wind.

Heat also plays a role in the formation of clouds, as it is heat which is responsible for evaporating (accelerating/vibrating water molecules in water until the point that they free from each other, or joined together only in small groups) water from the ground in the first place.

On a hot day clouds form because heat converts water on the ground into water vapor. Gravity still has a hold on water vapor, but the air molecules are pushing themselves to the Earth harder than the water molecules.

To clarify let's pretend that air molecules are large steel balls, and water molecules are somewhat smaller plastic balls.

The heavier steel balls work their way to the ground due to vibration and other factors (wind, heat, etc.) causing the plastic balls to rise, since the space on the ground is occupied by the steel balls.

Clouds form because the high micro-gravity of the water molecules draws them together, however, the water molecules do not form drips of water as they are vibrating/moving to quickly (due to heat) to be in a close enough proximity of each other to join.

Later on in the day, or at night, when the temperature begins to drop, the clouds may thicken.

As the clouds thicken and the heat decreases it becomes possible for the micro-gravity of the water molecules to push them together forming drips of water which fall as rain.

Also as the clouds thicken more micro-gravity is in one area, and may draw (lightning) electrons from the ground. Note: If a cloud becomes thick enough during the day when it is hot (temperature wise) outside, the cloud may block out enough heat to allow water molecules located in its lower portions to push themselves together and form drips of rain.

Either way, once the micro-gravity has formed and is holding a drip of rain together, at that point gravity then causes the water molecules to push themselves towards the Earth (as a drip of rain); but at the same time the micro-gravity in the drip of rain is causing the water molecules to push themselves towards each other, which maintains their proximity to each other thus maintaining their form as a drip.

In addition to micro-gravity holding together matter that we as humans can physically touch; it also holds/stores particles of matter known as energy, which we cannot physically touch. Some examples of energy held by micro-gravity are: photons of light, electricity, and radiation.

As mentioned earlier, some micro-gravity groups itself together, even without attracting or touching another physical object. For example, sometimes micro-gravity may group itself together in the air.

If you happen to be watching you may see a red dot appear for a brief moment during a grouped micro-gravity explosion (this is because micro-gravity is able to hold some types of electromagnetic radiation) which will be further discussed later on in this section.

Similarly if you are in a dark room and an area of grouped micro-gravity comes apart you may see a short localized flash of white light.

Many things can cause grouped/ungrouped micro-gravity to leave an area or object, including radio signals, heat, and impact.

An easy example of heat/impact causing a temporary loss of micro-gravity would be; if you were to heat or drop a magnet it would lose some of/all of its strength. Another example of impact causing loss of micro-gravity would be the scientific rule regarding force of impact. This rule states that as speed doubles force of impact increases by 4.

A clear example of this law in action would be a car accident. Let's say you drove a car into a brick wall doing 50 mph, there would be a strong impact evident by the damage to the vehicle and possibly the wall.

Let's say you drove a car into a brick wall at 100 mph, than there would actually be 4 times as much damage to the vehicle and the wall, as when you did it at 50 mph.

There are many other examples, so the rule, "as speed doubles force of impact increases by 4" is widely accepted and is used when designing cars, airplanes, engines, and weapons etc.

Although it is a widely accepted rule that appears to be true, it is not. Let me explain, you see when speed doubles force of impact does not increase by 4, but rather during the impact the physical strength of the objects impacting decreases by 50%.

The reason the strength of the objects decrease by 50% from their normal strength during an impact, is because the force of impact not only affects the objects that are impacting each other but also the micro-gravity which holds them together. This is because micro-gravity has mass too.

In summary, as the combined speed of the objects colliding doubles the force of impact doubles; during the impact the strength of the objects involved decreases by 50%. So basically it does appear as if when speed doubles force of impact increases by 4, but as mentioned above this is truly not the case.

For example a glass/cup is being held together because micro-gravity is causing the glass molecules to push themselves together, while gravity is causing these same molecules to push themselves towards the Earth. If you drop the glass and it impacts the ground, during the impact the glass itself becomes weaker as some of the micro-gravity temporarily changes direction.

When micro-gravity is not being emitted evenly from all atoms in all directions, the structures of the objects impacting become unstable, and thus weaker. As the speed of impact doubles, more micro-gravity is moving in the direction that the objects impacted in, thus the object is twice as weak during impact.

Again this is because force of impact not only affects the object itself, but also the micro-gravity which holds it together.

The last example of force or energy which affects grouped or ungrouped micro-gravity is highly debatable; however I believe it to be true. What I am referring to is electrically caused micro-gravity explosions.

Micro-gravity explosions are responsible for things/objects falling over by themselves, and also for things/objects floating into the air. Currently people explain away these events as ghost, poltergeist, PKE (paranormal kinetic energy), telekinesis, psychokinesis, psychic kinetic energy (PKE), psychic energy (PE), etc.

These explanations in my opinion are unfounded. The truth of the matter is that all these events are caused almost entirely by micro-gravity explosions.

Micro-gravity Explosions

What is a micro-gravity explosion?

A micro-gravity explosion is when a free floating electron shard smacks directly into a micro-gravity particle and splits it into millions of pieces. These broken micro-gravity fragments accelerate via dispersion then smash into other micro-gravity particles and brake them. Then the freshly broken micro-gravity fragments go on to brake even more micro-gravity particles, etc.

The release of billions of micro-gravity particle fragments in all directions may cause an object to levitate/float off of the ground due to pressure. This is because there are no particles to smash the broken micro-gravity particles (micro-gravity fragments) and there are so many micro-gravity fragments being released in all directions (at 187,000 miles per second), that there isn't sufficient time or space for the fragments to escape fast enough. So some of the fragments will stack on top of each other (so to speak) for a short time, and thus lift the object emitting them.

What happens to the object emitting these micro-gravity fragments?

There is often more then enough micro-gravity to keep an object intact (keep the atoms in place) during a micro-gravity explosion, depending on the mass and volume of the atoms the object is made of.

During a micro-gravity explosion the object emitting the micro-gravity fragments will certainly become weaker (easier to damage) and somewhat molecularly unstable. The structure of the molecules could change slightly in some parts of the object, however a light weight object will usually not become molecularly unstable to any appreciable extent, as with the case of a sponge, piece of cardboard, foam, rubber, solid plastic cup etc.

In the event when a micro-gravity explosion occurs with a more complex or heavier object it is actually quite likely that significant damage to the object will occur, especially if it was physically stressed, or if it hit another object while it was floating.

Also in a heavier or more complex object there may not be enough undamaged/non-shattered micro-gravity to hold the object together on a molecular level, meaning that some of the atoms that each molecule is made up of may change order, or escape the object entirely.

Additionally many of the molecules that the object is made up of, may become weaker or altered as the atoms in the object become free floating and start to escape. (A molecule is two or more atoms together, for example a water molecule is two hydrogen atoms and one oxygen atom, and is known as H_2O or HOH)

Again micro-gravity explosions are rarely observed, and when they are, they are generally attributed to something else such as poltergeist or psychokinetic energy.

Grouped micro-gravity frequently explodes, however these explosions are silent and usually smaller than 1 cm. (2/5 of an inch)

When a grouped micro-gravity explosion occurs in front of, or even inside of an electronic device; the device may start, stop, or function incorrectly. This is because grouped micro-gravity stores as much electromagnetic radiation as possible, then when a grouped micro-gravity explosion occurs that same amount of electromagnetic radiation is released.

Even grouped micro-gravity explosions are explained away as something else. For example if someone's TV turned itself on! That person may think that someone else turned it on, or that they had forgot it on, or that there was a power surge or a problem with the TV that caused it to come on, but more likely that person may simply not even care.

Now that I have explained what micro-gravity is and does, and what a micro-gravity explosion is; you may be wondering has anyone ever caught one on film.

Answer: Yes. As I mentioned earlier micro-gravity explosions are caused when a free floating electron shard splits a micro-gravity particle into fragments. A scientist by the name of David Hutchison has created multiple micro-gravity explosions and has captured some of them on tape/video.

Although he can't explain what he does, and can't really control the micro-gravity explosions that are caused by his manipulation of the equipment in his lab. What he does has been termed "The Hutchison Effect."

What does he do? Essentially he transmits radio waves at an object(s) from multiple directions. The power of the radio transmitters he uses is so great, that when a fluorescent light is placed near one of them it lights up, just from the energy of the radio signals.

After changing frequencies and voltages constantly (although this has little to do with it), eventually a micro-gravity explosion occurs and objects start to levitate. Effectively "The Hutchison Effect" is a micro-gravity explosion on video.

In summary micro-gravity is what causes atoms to hold themselves together and group near each other (forming molecules), while gravity causes these same grouped atoms (molecules) to push themselves towards the Earth. Ultra-micro-gravity is simply particles which are released by some types of micro-gravity which causes it to group, thus the reason we have grouped and ungrouped micro-gravity.

Grouped micro-gravity is responsible for such things as static electricity, and makes things such as magnetism possible.

Micro-gravity is responsible for things such as lightning, and holding atomic structures together.

Grouped micro-gravity explosions are responsible for starting, stopping, or temporarily changing the function of electronic devices. Also these explosions, although they make no noise, are often responsible for knocking over stationary objects, especially dense objects such as metal.

Micro-gravity explosions are possible and have been documented on video as "The Hutchison Effect." Micro-gravity explosions cause heavy objects to levitate and can also cause changes/damage to these objects on a molecular level when the molecules exchange atoms, or when the atoms simply escape.

In conclusion without micro-gravity there would be no life in our dimension, as all atomic building blocks would become free floating and form a single dull lifeless substance. Micro-gravity is all around us, though few people are aware of it, or even know about it for that matter.

Thank you for reading my thoughts, theories, comparisons, and explanations of gravity, micro-gravity, and ultra-micro-gravity. I truly hope you understood and enjoyed them, but if not, at the very least they should make you think.

Viruses and One Cell Organisms

My thoughts on viruses are plain and simple, "Viruses are alive!"

Since I started reading medical documentation I have discovered that doctors are trained to believe that viruses are not alive. My definition of life is that if something can consume material (food), produce waste, replicate, and die; than it is alive. An example of this would be that a flower or plant is alive.

My definition of intelligent life is very similar. If something can consume material (food), produce waste, replicate, and do things actively for a reason or purpose; and of course die, than it is intelligent life. An example of this would be a bee or an ant.

Despite the size of a virus it meets my definition of intelligent life regardless of the fact that for the most part viruses are parasitic in nature. They prefer to be guest in a host's body such as yours, regardless of the fact that they are unwelcome by the immune system. Antibodies are also forms of intelligent life, as antibodies do things with reason and purpose.

One of the reasons scientist believe viruses are not alive is because the protein chains that they consist of are too small to form a cell (so they say).

I have some news for you. What we call one celled organisms, e.g. protozoa, actually consists of many billions of cells. For example if a protozoa gets a cut it heals. For something to heal the formation of new cells is required. For something to have external body parts (e.g. arms and legs) more then one cell is required.

Since the cells are so small we can see right through the creature under the amount of light that we consider normal; and because we can see through the creature we consider it to be a simple one celled organism, although it is not.

If an enormous creature were to study us under a microscope using 10,000 times the amount of light that we consider as normal, he may believe that we are just simple one celled organisms with two arms, two legs, a head, and simple see through organs.

Just because you can't see the cells doesn't mean they're not there. Just because it looks simple doesn't mean it is.

Light and Vision.
(What light is! How we see it.)

What is light?

Some people think of light as a wave, while others think of it as a particle. It is understandable why people would think of light as a wave, as essentially particles of light do bounce off objects in the same sort of way a wave would.

Light is made up of particles called photons, however, even though light is comprised of tiny particles called photons it can still be thought of as a wave, just like an actual wave of water is made up of tiny particles called water molecules, where the wave is the motion of these molecules.

Since photons of light are always moving and expanding or spreading out, they do by definition form waves; so to properly answer the question, "What is light?" I'd have to say that light is three-dimensional waves made up of tiny particles called photons.

What is a photon?

Answer: A photon is an electron fragment. When an electron is smashed it forms photons. Photons come in all different sizes (spectrums). Not all spectrums of light, and in fact very few, are visible to humans. Most people are only familiar with the spectrum of light that we as humans can see. Our spectrum is known as the visible spectrum.

Note: Some spectrums of light do not pass through glass, while others pass through solid objects such as plastic. Just because a recycle bin is opaque blue in the visible spectrum doesn't mean it is in all spectrums, in others it could be as clear as glass.

Light is also called the electro magnetic spectrum, or electromagnetic radiation. The seven main categories/classifications of light in the electromagnetic spectrum are as follows; radio, microwave, infrared, visible, ultraviolet, x-rays, and gamma rays. We will briefly touch on each of these categories/classifications of light in more detail, a little bit later. First let's examine the flaws, inconsistencies, and errors in the current theories of today regarding light.

Currently when trying to understand light scientist look at the "length of the wave of light" or wavelength; and the number of waves that travel through a point of space over a given time (usually 1 second), this is known as the frequency. As mentioned at the start of this section photons of light are nothing more than smashed electrons, scientist measure the energy in each spectrum of light in EV or electron-volts.

Current scientific theories demonstrate/postulate photons (particles of light) may be created by energizing atoms; this is truly not the case. Photons are nothing more than fragments of smashed electrons that are created when two or more electrons collide. That being the case of incandescent, fluorescent, and sodium light bulbs, etc. which all produce light in this way.

Another way to accelerate electrons to the point of collision is by increasing the temperature by adding heat. Some examples are a lantern, woodstove, or fire, where combustion (combining the air molecules with other molecules e.g. gas or wood) cause electrons to smash into each other as the molecules combine. This is because when the molecules combine to form new atomic structures, e.g.

carbon monoxide, there is not enough free space around the atoms for all the electrons to orbit. Also simply the fact that heat is raising the temperature of the material being burned (causing an acceleration in the motion of the molecules/atoms) increases the possibility/amount of electron collisions.

Based on what was just mentioned it is not the energizing of atoms by adding heat, or the acceleration of electrons (as scientists currently believe) that creates photons (particles of light); but rather it is the electron collisions themselves that create photons, which are just smashed electrons.

The next point that I would like to make is that, although photons of light do expand and form three-dimensional waves, which in affect do limit the size of something which we can see (must be larger than the wave); it is not the wave itself that we see, but rather the photons; which we will discuss in further detail before the end of this section.

How are scientists measuring the energy (EV) of the different regions (spectrums) of electromagnetic radiation (light)?

Answer: Ultraviolet, x-ray, and gamma rays, are the 3 final regions of electromagnetic radiation and scientist measure their energy in EV (electron volts) as opposed to using wavelengths because the wavelengths of these forms of light are too small to accurately measure, or even think about for that matter.

The seven regions of the electromagnetic spectrum are as follows: radio, microwave, infrared (heat), visible light, UV (ultraviolet radiation), x-rays, gamma rays.

Let's briefly discuss each of them below:

Radio: This is a form of light because the radio signals that are transmitted and received by radio stations, TV stations, cell phones, cordless phones etc. are all comprised of shattered electrons.

Microwaves: This is a form of light (although not visible to humans) because again they are comprised of shattered electrons.

Note: Microwave radiation is used in devices such as the microwaves we cook our food with, also by microwave antennas or dishes that large corporations use to transmit and received data, or for sharing data access between nearby locations.

E.g. One large building is equipped with high speed Internet and shares its connection with another building 5 miles away via microwave antennas (microwave transmitter/receiver dishes). Aircraft and police radars also use microwave electromagnetic radiation.

Visible: This is the form of electromagnetic radiation (light) that we as humans can see. Visible light (electromagnetic radiation) is emitted by everything from the Sun (stars), to light bulbs, to televisions, to fire flies, to glow sticks etc.

Infrared: Infrared light is heat; objects exposed to something emitting infrared light (heat) gain temperature (increased molecular movement) and become warm or hot.

An object with a high enough temperature will produce/release some heat (infrared) due to temporary changes in the variable fixed proximities of the molecules leading to electrons colliding and becoming shattered. E.g. If you use a torch to heat a horse shoe to a glowing hot temperature, that horse shoe is then emitting infrared light/heat and can cause objects near it, which are not touching it, to become warm.

As discussed in previous sections of this book heat and temperature are different. Heat is smashed electrons (photons) as is light and can travel through a vacuum. Temperature is the vibration of molecules, and temperature transfer requires physical contact. E.g. You feel warmth when you hold a hot coffee cup as the vibrating molecules of the coffee cup are hitting the molecules in your hand, so to speak, and causing them to vibrate.

Infrared is used in electronic devices such as TV remotes and infrared ovens. Infrared detectors are used by police and military personnel to track people in the dark, as human beings (as do all animals) emit some heat or infrared, although only at very low levels.

Ultraviolet: Our Sun is a source of UV as are all stars. UV rays cause sunburns. There are some industrial beneficial uses of UV such as the curing of inks and high-gloss varnishes, industrial coatings and finishes on wood, metal and plastic products, and the curing of adhesives for bonding electronic components. Also it is sometimes used by water filtration plants to kill bacteria without the use of harsh additives such as bleach (chlorine).

X-rays: Can penetrate many solid objects that visible light cannot, and are used in x-ray machines at the doctors and dentists to take pictures of your bones and teeth. X-rays are naturally emitted by neutron stars and hot gases in space.

Gamma rays: High-energy electromagnetic radiation like x-rays, but of a much higher energy, frequency, and shorter wavelength. Gamma rays occur naturally in our universe and often hit the Earth in short-lived bursts of a few seconds (this occurs daily).

Gamma rays are also produced on Earth from the radioactive decay of natural and man-made materials, such as the fuel rods used in nuclear power plants. Radioactive decay will be discussed further in the next section of this book entitled "Nuclear Reactions and Radiation."

Now that we have discussed what light is, the production of light, and the various spectrums of light, let's take a look at how we actually see with it.

Vision

What is vision?

Obviously vision is our ability to see light, but how does it actually work?

The ability to see is dependent upon the function of several structures in and around the eye. Note: We will not be covering the anatomy of the eye but we will touch briefly on the function of each piece as needed.

The miracle of vision begins with rays of light reflecting from the object or objects that you are looking at to the cornea of your eye. The cornea is a transparent, dome-shaped window covering the front of the eye. It is half a millimeter thick and consists of five layers which we will not be touching on. For the purpose of explaining vision you may think of the cornea as the outermost transparent surface of the eye.

The rays of light are then bent, refracted, and focused by the cornea (discussed above), lens, and virtuous.

Before continuing let's explore what the lens and virtuous are/do. The crystalline lens is located behind the iris which is the colored part of the eye that is visible from the outside. E.g. Mark has blue eyes, and Sarah has brown eyes. The iris is actually a circular set of muscles which control light levels inside the eye, similar to the aperture of a camera. The purpose of the lens is to focus light directly on to the retina.

The virtuous is a thick transparent viscous substance (similar to that of egg white). It is comprised mainly of water and makes up about 65% of the eye's volume. The virtuous helps give the eye its round shape. Also its viscous properties allow the eye to return to normal shape if compressed.

Let's recap and progress on what we just discussed. When you look at an object or objects the rays of light which are reflecting from those object(s) to your cornea are then bent, refracted, and focused by the cornea, lens, and virtuous. The purpose of the lens is to ensure that the light comes into sharp focus on the retina. (Note: The resulting image on the retina is upside down.) The retina converts the light to usable electrical signals which travel up the optic nerve to the brain.

In a moment we will discuss exactly how the light is converted to usable electrical signals but first let's thoroughly examined the different components which make up the retina.

The retina is comprised of a multi-layered sensory tissue that lines the back of the eye. This tissue which the retina is made of contains millions of photo-receptors that capture light rays and convert them into usable electrical signals.

The two types of photo-receptors within this photo-receptive tissue (the retina) are rods and cones. The retina of each eye has approximately 7 million cones which are contained in the macula, which is the part of the retina that provides us with central vision. Cones utilize bright light, provide us with sharp central vision, and allow us to have color vision.

In addition to the 7 million cones (as mentioned above), the tissue which the retina is made of also contains approximately 125 million rods which are spread throughout the rest of it. Rods function best in dim lighting and allow us to have both peripheral vision, as well as good night vision.

Note: Both cones and rods can be damaged by excessively bright light, or sharp light such as ultraviolet, metallic (e.g. blue metallic headlights on some cars), blue light (e.g. energy-saving light bulbs).

We will be fully discussing the complexities of the different types of colors that our eyes are able to perceive. But first let's recap on what we've covered so far.

So far we've discussed how visible light bounces off the objects which we are looking at and enters our eyes, then those rays of light are bent, refracted, and focused by the cornea, lens, and virtuous.

The lens focuses light directly onto the retina. The virtuous is a clear gel like substance which gives our eyes their shape and allows them to recover from injury. Light travels from an object that we are viewing to the cornea, and then is focused by the lens directly on to the retina.

In order for light to reach the retina it must travel from the lens, through the vitreous (full name: vitreous humor). Once those rays of light reach the retina they are organized into electrical signals/impulses which then travel through the optic nerve to our brain.

So just how does light get converted into electrical impulses that the brain can translate into images?

To answer that first we must remember that although light acts as a ray or waive it's truly a particle; and a particle of light (photon) is nothing more than a fragment of a shattered electron.

Because of the nature of shattering, each particle of light is of a slightly different size and shape from one another; much the way shards of glass would be of different shapes and sizes if you were to shatter a window with a baseball bat, as an example.

Before we can answer the question of how particles of light ("photons", which are fragments of electrons) get converted into electrical signals that our brains can convert into images that we can see, we must first further examine the function of the rods and cones in the retina of our eyes.

The retina of a human eye contains approximately 125 million rods. These 125 million rods come in approximately five different densities, or in other words consist of approximately five different masses. In reality thousands of different masses, but they fall into five different ranges. E.g. 9000 to 10,000; 100,000 to 120,000; 900,000 to 1.1 million; 3.2 million 3.3 million; 4.8 million to 5 million: although the rods are all very similar in size.

The difference in mass is caused by the construction of the rods. Some rods have a narrow central chamber and outer veins that ultimately lead to the center of the rod, which than connects to the optic nerve.

The liquid which makes up for most of the mass in each rod, and that is held in the rods outer veins and central chamber is called retinal, or visual purple. It has been nicknamed visual purple because of its purplish appearance when exposed to light.

Visual purple is a highly complex structure, which although we will not be covering what it's made up of, we will be covering exactly how it performs its function and precisely what that function is.

To recap a bit we have just discussed how rods are all roughly the same size, but can be split up into five groups of masses. The masses are different based on the amount of retinal (visual purple) in each rod due to the different structures of the rods. The rod structures vary based on the amount and size of the outer veins, and the size of the inner central chamber which connects to the optic nerve.

We have also discussed how light is nothing more than shattered electrons (photons). The stream of photons enters the eye through the cornea and is then bent, refracted, and focused by the cornea, lens, and virtuous.

What we haven't yet talked about is the role of visual purple/retinal. Retinal contains an extreme amount of micro-gravity (more than water, even more than acid); and photons (particles of light) are nothing more than shattered electrons. The rods with the highest micro-gravity (level 5) attract photons from the smallest in size to the largest, whereas the rods with the smallest mass generally only attract the smallest photons. This is because the largest photons are more attracted to the rods with the largest mass. The only photons left for the rods of low mass to attract are the small photons.

The largest photons are tiny in comparison to any of the rods. One rod can hold up to a million photons. It's a sliding scale, level 5 rods pickup on/attract the largest to the smallest photons of visible light. Level 4 rods will pickup on all the photons of visible light except for that which the level 5 rods already did. Level 3 rods will pickup on all the photons of visible light except for that which level 5 and 4 already did. Level 2 rods will pickup on all the photons of visible light except for that which level 5, 4, and 3 already did. Whereas level 1 rods will usually just pickup on the smallest photons of visible light.

When I use the term attract I am speaking metaphorically of course, the photons of light push themselves towards the greatest source of micro-gravity much the way electrons do, and rightly so, after all photons were once electrons themselves, so it's not surprising that they share many of the same characteristics such as the speed at which they travel (the dimensional limitation of 187,000 miles per second).

This dimensional limitation will be further discussed in the "Dimensional Sciences" section of this book, in addition to the force (micro-gravity) which contains/attracts photons, as well as what accelerates both electrons and photons to 187,000 miles per second (max speed in our dimension, dimensional limitation).

Please note that objects never really absorb light as one would have you believe but rather they further smash the photons making the photons smaller, which then reflect back to our eyes. Some of the photons become so small that they actually get caught/stored in objects and are released slowly in the infrared spectrum of light (heat), in addition to some of the other spectrums such as radio waves etc., depending on what spectrum the source of light was originally from.

Let's recap a bit and get back on track. The light from a light source hits an object(s) which is said to convert and absorb some of the photons, and convert & reflect some of the photons to our eyes, where they are focused by the cornea, virtuous, and lens onto the retina of our eyes.

Note: The virtuous humor and aqueous humor have an additional function which is slowing the photons of light down from 187,000 miles per second to less than 500 miles per second. It is the crystalline structure of the aqueous humor and virtuous humor which are responsible for this great feat.

Next the slow-moving (500 miles/sec) photons push themselves to their respective rods as discussed earlier. There is a lot of micro-gravity in the veins of the rods, but there is more in the inner central chamber of the rod as there is more retinal.

The photons then push themselves into the central chamber of the rod as there is more available micro-gravity there than in the veins. The available micro-gravity in the outer veins of the rods is being decreased/consumed by the very presence of the photons that they are absorbing.

Once in the central chamber/core of the retina the photons (shattered electrons) group together to form electrons. This occurs because the photons which are emitting micro-gravity are now free to push themselves together because other outside forces (e.g. Micro-gravity of skin and tissue cells) are barely affecting them, do to the fact that they are surrounded by an almost even amount of micro-gravity while they are in the retinal.

The larger the electrons become, the quicker they travel towards the brain via the optic nerve, or in other words once the photons form electrons and as the electrons reach there full-size they accelerate up to higher speeds as they travel to the brain; and this is because they are no longer being slowed down by the aqueous or virtuous humor.

Also they are pushing themselves towards the brain as fast as possible as they have the full mass and micro-gravity of electrons. Electrons (electricity) flow(s) from the greatest source of electrons to the greatest source of micro-gravity using the easiest path available; in this case the path is the optic nerve. The brain contains huge amounts of micro-gravity.

These electrons travel into the brain from the various 250 million rods (125 Million per eye) and are present in the visual cortex (simply the first place to receive and accumulate the electrons as a complete image) on the back of the brain for a split-second. From there they travel in pulses to our short-term memory at around 240 frames per second. (1 frame = a group of up to 14 million electrons in sequence.)

Our conscious receives/intercepts about 40 of these frames. The other 200 frames are channeled in pulses into long-term memory where they are visible only by our subconscious. If the subconscious shares access to these images with the conscious, than a person would be said to have what is known as a photographic memory. Note: The frames we see as images are not stored in long-term memory and are ultimately lost. All the above happens over the course of one second.

What we just discussed is known as gray-scale vision (non color). Your probably wondering how the different shades of gray are formed.

The answer to that is simple, remember how level 1 rods only pickup on the smallest photons, and level 2 pickup on those same size photons plus larger ones, and level 3 picks up on all those that 1 and 2 do plus larger ones etc. Effectively level 1 rods pickup on smaller photons than levels 2, 3, 4, and 5, respectively; objects reflecting smaller photons appear as darker objects to our brain than objects reflecting the larger photons.

This is because it takes around 300 times more of the smaller photons to form an electron than it does larger ones. The lack of electrons is the lack of brightness for our brains to work with, thus various parts of the image will appear darker than others depending on the size of photons being reflected to our eyes from the objects we are viewing.

Gray-scale vision requires less light than color vision which we will be discussing in a moment, but first let's take a look at where 3D vision comes from. Our conscious is able to visualize flat images as

3D for a variety of reasons which are purely mental, some examples below.

Ex. 1) Occlusion: If an object is covered by another object in our field of vision, than we may infer/perceive that the object is behind the other object, providing that we are aware that there are actually two objects and not one. This determination of the number of objects being viewed is achieved by viewing a set/pair of objects from more than one angle, for example 90° and 80°.

Ex. 2) Shading or Shadowing: If we know the location of a source of light, than based on the number of shadows, location of shadows and brightness, we're able to determine and separate objects, distance, and ultimately view them as 3D. As an example let's say there is a lamp and two boxes. The lamp is beside the first box and the second box is located directly behind the first box.

From different angles you can definitely tell that there are two boxes, and that one is behind the other. If you look directly at the first box from directly in front of it, it may be hard to tell that there are two boxes; however, we still know there are because of the locations of the two shadows.

Further more let's say there is a room illuminated by a single light. We can judge distance based on brightness, being able to distinguish distance is helpful when determining an objects size.

Ex. 3) Geometric perspective: As you know, parallel lines appear to converge as they go off into the distance, like looking down a long straight highway, the two edges of the road appeared to meet off in the distance. Your conscious mind compares the two lines of sight, from your two eyes, to judge distance and size.

Knowledge can also be used for judging size and distance. For example, since you already know how large a soccer ball is, based on its perceived size you can tell the distance, much the way you judge the distance in a picture.

There are other methods of perceiving things as 3D; generally, with the exception of a picture more than one method of visualizing 3D is used. Even when watching a movie or TV you are using occlusion, shading, or shadowing, and because objects are constantly being shown on slightly different angles you use angular perspective to determine the distance and size of the objects.

Let's go over what we have covered so far. By now you should understand how your brain is able to perceive different brightnesses of light, how gray-scale vision works (Gray-scale vision is also known as night vision as it works with far less light than color vision.), you should also have an idea of how you consciously perceive 2D images as 3D (mainly from seeing the same set of objects from multiple angles).

3D vision is a mental thing. If a person were to lack the conscious ability to judge size and distance, based on viewing images on various angles, than all images would appear 2D; and the only way that person would be able to distinguish objects from each other and judge distance would be from their own personal knowledge of the size and shapes of the objects. With enough mental deficiency a person would see the world as if it was a picture (flat/non three-dimensional).

There have been a lot of rumors that women see differently than men; they are for the most part not true. The eyes of men and women work in exactly the same way, and women are able to perceive the world as 3D just as easily as men can.

Color Vision

This actually works very similar to gray-scale vision with a few main differences. In addition to using 125 million rod shaped light receptors that convert photons of light into electrical signals, your eye is also using 7 million cones which are concentrated in the macula, especially in the fovea (the very center portion of the macula).

Similar to rods, cones are all roughly the same size, and also have different masses, but instead of there being 5 different groups of masses for cones there are actually 15, and this is despite the fact that there are only approximately 7 million cones in each of our eyes.

Cones are responsible for providing sharp central vision and color vision. We only need 7 million cones because as mentioned a moment ago they are most densely packed within the fovea, which is the central part of the macula, and the macula receives more light than any other part of the eye because light is concentrated there by the cornea, virtuous, and lens.

Color vision is distinctly different from gray-scale. When a photon of white light hits an object, or the outermost molecules (surface molecules) of that object (e.g. paint), it shatters. The size of the photons that are created is dependent upon the hundreds of different molecules they hit, as well as the density of the array of molecules within the outer surface of that object.

If the molecules are in closer proximity of each other than they are being held in place harder by micro-gravity, so when photons hit the molecules, the molecules will not move as much, thus they provide less cushioning and therefore the photons shatter into smaller photons.

With color vision there are five different classifications of masses of the cones in our eyes for each of the primary colors, which work similar to gray-scale. There are five levels for red, five for yellow, and five for blue, but it is a sliding scale. The yellow light receptive cones have the smallest mass, the blue light receptive cones have a midrange mass, and the red light receptive cones have the largest mass.

A level 5 yellow photo receptive cone picks up on level 5 yellow photons and all smaller photons (yellow levels 4, 3, 2, & 1). A level 4 yellow photo receptive cone picks up on all yellow photons except level 5. A level 3 yellow photo receptive cone picks up on all yellow photons except levels 5 and 4. A level 2 yellow light receptive cone picks up on level 2 and 1 yellow photons. A level 1 yellow cone only picks up on the smallest of yellow photons. A level 1 yellow cone produces the darkest/dimmest shade of yellow we can see, whereas a level 5 yellow light receptive cone produces the brightest.

The less photons your eyes receive the darker the overall image, and the more photons the brighter the image that you see. It can be a bright sunny day and you are still able to see a dull yellow object, you can see it better as it's brighter. It can be late evening when the Sun is going down and you could see a very light shade of yellow on a sports car driving down the road, the image looks darker overall because there is a lower quantity of photons entering your eyes, and in this example specifically, a lower quantity of yellow photons; however the yellow photons entering your eye from the surface molecules of the car's paint are level 5 (the brightest/largest yellow photons), and the large quantity of yellow photons you saw when you were looking at the dull yellow object in the bright Sun were level 1 yellow photons.

The overall brightness of the image you perceive is based on the total amount of photons entering your eyes, and the overall tint of the color is based on the mass of the photon, (level 5 photons have the highest mass and level 1 the lowest). Please try not to confuse tint with brightness. Tint is the brightness a of color based on a fixed amount of light.

You should now understand the difference between brightness and tint, in a moment we will be discussing shades (combinations of colors), then contrast.

Shades:

Now here's where things get really complicated. As you probably guessed your red and blue photo receptive cones have a higher mass than your yellow cones, but otherwise work in exactly the same way. This is true so there is one major thing to consider; a red photo receptive cone of any level will also pickup on all levels of blue and yellow photons; and a blue photo receptive cone of any level will also pickup on all levels of yellow photons, but no red ones; whereas yellow photo receptive cones only pickup on yellow photons.

Level 5 red photons are the largest and have the highest mass, whereas level 1 yellow photons are the smallest and have the lowest mass. There are 15 total mass classifications of color photons in 3 groups; red, blue and yellow. Five levels per color.

Like with rods, cones have veins and a central core, and because they contain different amounts of retinal they all have different masses. As mentioned a moment ago there are 15 different masses of cones which can be divided into three groups as shown below: red L5, L4, L3, L2, L1; blue L5, L4, L3, L2, L1; yellow L5, L4, L3, L2, L1.

A level 5 red cone has the highest mass, whereas a level 1 yellow cone the lowest. This is why a level 5 red cone will pickup on all photons. Your brain knows this so let's say you are looking at a yellow object, although the red and blue cones are also delivering a signal with the red cones signal being the strongest, the blue cones signal the next strongest, and the yellow cones signal the weakest, your brain will still show the object as yellow.

This is because your brain received a small but proportional amount of yellow signal, or in other words your brain knows that the object is yellow because it is aware that if you were actually looking at a red or blue object it would receive little or no yellow signal (depending on the shade of red or blue, which could contain some yellow). Since your brain has received some yellow signal in proportion to the red and blue it is able to determine that you are looking at a yellow object.

Your brain is wired/aware of the way that the cones in your eyes work and it uses proportions to determine/display colors correctly. If you were looking at a blue object, there would be a lot of signal from your red cones, some signal from your blue cones, and little or no signal from your yellow cones; because there is proportionately some blue signal your brain will show you the image as blue. If you were looking at an absolutely red object you're blue and yellow cones would send out little or no signal.

When your brain determines that an object you are looking at is reflecting both red and yellow photons at the same time it averages the signals together to form orange. When your brain receives enough blue and yellow signals from the respective cones at the same time, from an object or area of an object that your eyes are focused on, it averages it to be green.

The three primary colors are red, yellow, and blue. I state this because the cones in our eyes pickup on red, blue, and yellow photons of visible light. But what about RBG (red, blue, green) like on a television or computer screen? How can you make yellow from red, blue, and green since yellow is one of three primary colors?

Answer: Remember how we discussed that there are three classifications of cones each with five levels (for brightness), and remember how when you look at a true red object your red cones send a signal, when you look at blue objects both your red and blue cones pickup on the blue photons of light and thus both the red and blue cones send a signal, and when you look at a yellow object all three cones send signals to your brain.

Your brain is used to receiving signals from all three classes of cones when receiving yellow light, and it determines the shade of yellow via proportion. Therefore if you project a red light and a green light on a surface such as a movie screen, or TV screen the resulting color you will see is yellow as your brain has effectively received the same signals from the cones as if you were looking at something that was actually yellow, which in fact you were.

When red light is projected your red cones send a signal to your brain, and when green light (yellow and blue) is projected into the same area you effectively have red, yellow, and blue light entering your eyes from the exact same spot. Your brain knows that when you look at something yellow all three classes of cones will send it a signal, and that's why when you view a red and green (blue and yellow) light projected in the same spot you see yellow.

Note: There are no distinct photons of light in the visible spectrum that can be called Green. Green is simply a mix of blue and yellow photons. There are very few absolutes and your brain determines the color via proportion.

When you look at a yellow object your brain is receiving signals from all three classes of cones proportionally, meaning the red cone is sending the strongest signal, then blue next strongest, and yellow the weakest.

That yellow signal although weaker then the red and blue signals is very high in proportion to the stronger red and blue signals. If the object wasn't yellow than the yellow signal would not exist (certainly not to that extent) and if it was a color with only a tinge of yellow than the yellow signal would still be hundreds of times weaker. So although all three signals exist and yellow is the weakest of the three (but stronger than usual) your brain is able to determine that you are looking at a yellow object via proportion.

When you look at a green object your brain is still receiving signals from all three classes of cones, however it is receiving slightly more red signal, much more blue signal, and about half the yellow signal of that when you were looking at a yellow object; and by this your brain is able to determine via proportion that you are looking at a green object (blue and yellow in the same area at the same time).

Theoretically if you look at an absolute red object or source of light, only your red cones would send a signal to your brain, however in practicality nothing is 100%. When you look at a red object there's a million to one odds that it is reflecting back some blue and yellow photons as well, however most photons reflecting into your eyes from the object are red and your red cones would proportionally be sending a dominant red signal to your brain, and thus via proportion your brain is able to determine that the object you are looking at is red.

Do your cones allow you to see the colors black and white or just your rods?

Answer: When your eyes receive enough light to activate (get a signal from) your cones, your brain treats that signal with priority for the purpose of defining colors. That being said your brain is still using your rods as a tool of comparison to cross(double) check the signal it is getting from your cones for the purposes of defining shape, distance, brightness, and ultimately 3D vision.

Your cones allow you to see black and white. When you look at something that is white your yellow cones are receiving as many yellow photons of visible light as if you were looking at a yellow object, you're blue cones are receiving as many blue and yellow photons as if you were looking at a blue object, and your red cones are receiving as many red, blue, and yellow photons as if you were viewing a red object. Essentially your brain is receiving three dominant signals and knows the object you are looking at is reflecting all visible light proportionately and evenly in all base colors, the result we see is white.

When we look at a bright light again the same thing is happening only the signals are much stronger but the proportions are the same and we see white. White is an even amount of all 15 levels of visible light or, a proportionally even mixture of red, blue, and yellow light. Red photons are the largest, blue the next largest, and yellow the smallest. There are five levels for each of the three base colors, the size of the photon determines what it is. Photons themselves are nothing more than smashed electrons.

When you look at a black object your cones are sending the weakest possible signal to your brain, however the proportions of red, blue, and yellow are the same as when you look at a white object. Black, like white, is an even proportion of all three colors, only at a much lower level. A black object reflects red, blue, and yellow photons proportionately, only much less of them. Your brain sees this even but week signal as black. Black is a week even mix of photons, or no photons at all (darkness/absence of light).

Gray is also like white in the way that all 3 base colors are evenly proportioned. When you look at a gray object there is less light (evenly proportioned) reflecting back to your eyes then from a white object, but more than from a black object.

When the proportions are not even you end up with a tint, as an example let's say there was slightly more blue/yellow photons than red ones reflecting from an object; under those proportions depending on the total amount of light the object reflected, the color could be midnight black, off-white, or bluish gray. Again there are no absolutes, with three base colors and variable amounts of light the possibilities of different shades/mixtures of color are as infinite as the mathematical calculation for pie 3.14 . . . .

Why are there five levels of cones for each base color? Why 15 total different types of cones, five yellow, five blue, five red?

This is the way in which our eyes are designed, or have developed over time. Having five cones of different masses for each base color allows us to see colors very accurately.

Suppose we only had three types of cones one for red, one for blue, and one for yellow. Now let's say you were looking at a bright yellow object. First of all, one moment the object may look red, another moment green, and a moment later blue, another moment orange, and another moment yellow. This is because the smaller sized yellow photons may more easily be picked up by the blue cone, and especially the red cone.

Due to this lack of accuracy the color of an image could literally change just from a minor change in brightness or viewing angle, and at the very least the shade of a yellow object could literally change several times a second, that in itself would be unnoticeable as the color would average out between the visual frames, but the tint of the object could easily change from moment to moment.

Essentially having 15 total levels (5*3) of cones provides us with far (hundreds of times) more accurate color vision. Five is better than one!

Averaging is simple and is done with gray-scale and color vision. If you move your hand fast enough in front of your face it will appear to blur, this is not because of a lack of frames per second (240), or due to the speed at which light hits our retina's (500 miles/second), but rather because of a limitation in the speed at which our visual cortex can access our visual memory.

As an example when you watch TV you are looking at the same spot, where there are 30 pictures per second flashing in front of you. You can't tell that they are individual frames/pictures and the picture you are looking at seems to be fluid. If someone sets down a glass on a TV show, you are still using averaging to view it, as instead of seeing the 30 frames, you see motion.

Even a high speed car chase on TV seems like smooth motion, despite the fact that the car and especially its surroundings are in very different places each frame. This is because it takes time for the visual cortex (visual processing center) of your brain to access your visual memory.

If you take a white circular wheel and attach it to a drill and paint one-inch black stripes every second inch you have a black and white striped wheel. If you spin it, then it appears to be gray, this is again averaging (visual memory access rate limitation). The visual cortex in your brain updated the white lines to be black, and the black lines to be white, but they changed position before it completed, so it started updating the other way around, but the positions changed again before it could complete. Half white and half black is gray.

If a bright light flashes on and off 50 times a second, you might think that it's only half as bright, and would not notice the flicker. A bicolor light emitting diode produces bright green or dark red light depending on which way you attach the wires. Certain electronic devices such as cell phones use a bicolored LED (light emitting diode) to produce a third color (usually amber) simply by changing the direction in which the current flows through it 30 times per second. You don't see red, green, red, green etc. 30 times a second, instead the two colors are averaged and what you see is amber. This is because of the rate at which your visual cortex is able to access your visual memory.

Both the above examples demonstrate visual averaging.

Let's recap a bit: We have now covered light, gray-scale vision, 3D vision, color vision, visual averaging, and the way in which our eyes work.

There are many diseases of the eye which we will not be covering in detail. Diseases of the eye such as cataracts are caused by bacteria. Other physical problems with the eye such as nearsightedness or farsightedness are a result of weak inner eye muscles used to focus, or because of a hardening of the physical components due to lack of use, or from physical trauma to the eye.

Color blindness:

The second last item I would like to discuss in this section is color blindness. Not all animals are color blind, scientists and laboratory workers have via scientific method determined that certain animals (gorilla family) have full tricolor vision while others have bicolor vision meaning they see color but only combinations made from two base colors instead of three.

Amongst different species of animals with color vision, there may be up to four different ways to see an object of color. The same object may appear as four different colors to the various animals. 1. R-B-Y (tricolor), 2. B-Y (bicolor), 3. R-Y (bicolor), 4) R-B (bicolor).

As for color blindness in humans; either total color blindness, or even just partial such as in the case that one base color is missing, remains extremely rare. It is possible for a human to have a deficiency in one of the base colors, meaning that he/she doesn't see as much of it, but it would be very rare for someone to be totally color blind in one of the base colors.

There are many different types of color blindness all with different names but essentially most of them are referred to as red-green color blindness, meaning that a person (with red-green color blindness) is unable to fully distinguish one of the base colors during certain lighting conditions, or under certain waive lengths of light (size of the photon). E.g. Something green looks blue under certain lighting conditions.

Color blindness test:

Numerous people are miss-diagnosed as color blind when in fact they have completely normal color vision. There are flaws in both of the two main color blindness tests which I would like to make you aware of.

In the Ishihara color vision test, which consists of a series of pictures of colored dots with an embedded number, the viewer is required to correctly determine the majority of the numbers he/she is viewing.

This test in my opinion is far from accurate for the following reasons:

1) The numbers are often misshaped, as if they were written by a five year old child, making them harder to distinguish.

2) The numbers are not formed from a single base color, but rather the dots that the numbers are comprised of are made up of two base colors. E.g. The dots forming a number may be various shades of light orange, pink, and red, on a medium orange, green and yellow background.

That being said people who have normal color vision may not be able to see the various numbers, although they are distinguishing between all the colors on the card. This is because their brains are not used to making images out of random colors (hallucinate); whereas someone who cannot distinguish between all the different shades of orange, red, pink probably wont have any issues with seeing the numbers as most or all of the dots appear to be the same color. The Ishihara is the least accurate, most commonly used test. I suspect this test produces between 35% to 50% false positives each year.

Note: There are special color filter contacts and glasses available on the market today, most are custom-built. These lenses filter out certain ranges of the visible spectrum so that the person wearing them finds colors to be more vivid and distinct. This is because they see less of them! Nearly 100% of people who have previously failed the Ishihara test, and are labeled as color blind, pass when wearing these lenses.

Why would decreasing the number of visible colors one can see allow them to pass this test?

Answer: If the person who failed the test was mentally normal and had normal color vision, than that person would be unable to see the roughly shaped numbers comprised of various colors on a mixed color background. All they would see is 25 different colored dots of various sizes.

If instead the numbers appeared to be made of one color instead of five it would be possible/easy for that person to identify the numbers. People who pass the Ishihara test without color filter lenses are often either slightly color deficient or may have a slight mental illness such as schizophrenia. Please note that a person wearing color filter lenses cannot pass the Farnsworth D-15 color vision test.

The Farnsworth D-15 color vision test is better, but not perfect. In this test the subject is asked to arrange 15 colored one-inch chips according to color. This test is far more accurate than the Ishihara test as it does not require the subject to visualize numbers made of different colors, and therefore requires the subject to prove he/she can distinguish between various colors.

Also the D-15 does not allow for people who are in fact color deficient to pass as normal; whilst people taking the Ishihara test who are not color deficient, but who cannot see the numbers because their minds are not accustomed to putting together patterns made of various colors (which normally they shouldn't be) fail.

While this test is clearly better than the Ishihara test it's still not perfect. Since it only utilizes 15 chips of various colors, it can leave out minor color vision defects.

The ultimate color vision test

My idea for the ultimate color vision test is similar to the Farnsworth D-15 test, but instead of using just 15 color chips the subject would be given 90 color chips in a bucket (15 chips of various shades of each of the base colors; 15 red, 15 yellow, 15 blue, totaling 45 hues of the base colors). Also 45 chips of intermixed colors (15 orange, 15 green, 15 purple) which are colors created from mixing the base colors. The subject would be required to separate the colors red, orange, yellow, green, blue, purple, and then arrange each of the six piles of 15 color chips into order based on the shade of the color.

CCD chips versus the eye

A CCD chip is and electronic eye used in digital cameras, video cameras, web-cams etc. and is different from the human eye in almost every conceivable way.

The CCD chip (electronic eye) relies on photons to knock electrons out of a grid. A computer chip is then able to determine how much light was reflected to an area of the grid, by the number of electrons needed to refill/replace the missing electrons from that same area.

Our eye on the other hand, instead of taking electricity (to determine where light was), it combines the photons of light back into electrons which are then sent to our brain. This is done via a chemical called retinal. In short our eyes make electricity while a CCD chip uses electricity.

Also our eyes use different cones to send different color signals to our brains, thus our eyes use every bit of light they receive, whereas the CCD chip's electronic photo-receptors (grid of electrons) are all the same and color filters are used so that some photo-receptors receive only red light, some receive only green light, and some receive only blue light.

Note: There are twice as many green filters used, as there are red or blue.

Next to our brain our eyes are one of the most complex organs in our bodies.

It's amazing how humans have created the CCD chip without truly understanding light or vision, how we invented electrical generators without truly understanding magnetism or electricity, not to mention how we created computer chips without truly understanding logic or intelligence. What will we think of next?

I hope you enjoyed this section and have obtained a better understanding of what light is and how we see it.

In the next module we will be looking at the topic nuclear reactions and radiation which includes a subsection on fusion and fission. Essentially the next two parts of this book are devoted to nuclear science.

If you found this section interesting than I'm sure you will enjoy the next section of this book entitled "Nuclear Reactions and Radiation."

Nuclear Reactions and Radiation

In this section we will be discussing nuclear reactions and radiation. There are only two main reasons to create a nuclear reaction, one being a nuclear bomb, and the other being the generation of nuclear power. For our purposes we will mainly be looking at the generation of nuclear power.

Around 17% of the electricity used worldwide is generated by nuclear power plants. Some countries use it more than others. In France approximately 75% of all generated power is nuclear, and in the United States about 15%. The size of the country must be taken into account as well. There are over 400 nuclear power plants worldwide, and more than 100 of them are in the United States.

In a moment we will be looking at how a nuclear power plant works, but first we will be covering the various isotopes/substances commonly used in the creation of a nuclear reaction.

Uranium is a fairly common element on Earth and was incorporated into our planet during its formation. Uranium is formed by stars such as our Sun.

Uranium 238 has an extremely long half-life of approximately 4.5 billion years, meaning that every 4.5 billion years the amount of uranium present on our planet divides in half. Due to its extraordinarily long half-life it is still present in fairly large quantities today.

Uranium 238 or U238 makes up around 99% of the uranium found on our planet. U235 makes up about 0.7%. U234 is even rarer and is formed by the decay of U238.

U238 goes through many stages of decay to form a stable isotope (lead). Transforming into U234 is just one of the stages in the decay of U238.

Both U238 and U235 decay naturally via the release of alpha radiation. We will be discussing radiation shortly; however U235 has a unique property which makes it invaluable for both the production of nuclear power, and the creation of nuclear bombs.

U235 does not just undergo spontaneous fission like U238 or U234, but it can also undergo induced fission; meaning if a free neutron slams into the nucleus of a U235 atom, the U235 atom will split immediately creating two smaller atoms and releasing nuclear radiation as well as energy in the forms of heat and light. 2-3 neutrons are also released as the U235 atom splits. The two new smaller atoms emit gamma radiation as they settle into their new states.

Types of radiation:

Alpha radiation: Alpha radiation is a heavy, short-range particle, and is actually the ejected nucleus of a helium atom. Most alpha radiation is unable to penetrate our skin, however materials which emit alpha radiation can be harmful to humans or animals if inhaled, swallowed, or absorbed through open wounds.

Alpha radiation travels only a few inches through the air and is not able to penetrate clothing or a sheet of paper and therefore is not an extreme hazard.

There are a variety of instruments designed to measure alpha radiation, such as a thin window Geiger Muller probe or "GM probe" used almost everywhere here on Earth where radiation is a concern.

There is also the prospectors alpha particle spectrometer or "APS", which is used by NASA to detect alpha radiation in space.

These instruments cannot detect alpha radiation that is behind a thin layer of water, dust, or paper etc. because alpha radiation is not penetrating. Some examples of substances which emit alpha radiation are; radium, radon, uranium, thorium etc.

Beta radiation: Beta radiation is a lite weight, short-range particle, which is said to be an ejected electron. Beta radiation is able to travel several feet through the air and is mildly penetrating. It can penetrate human skin to the germinal layer, which is where new skin cells are produced.

If large levels of beta radiation emitting substances are left in contact with human skin for a prolonged period of time, obviously their harmful affects could cause skin injury or cancer. It goes without saying that ingesting or inhaling substances that emit beta radiation could be potentially harmful.

Clothing provides a minimal amount of protection from beta radiation. Most beta radiation emitting substances can be detected with a survey instrument, or a pancake type GM probe. Some examples of substances which emit beta radiation are; strontium 90, carbon-14, tritium, and sulfur 35.

Gamma and x-ray radiation:

Gamma radiation and x-rays are both able to travel through miles of air, and several inches through the human body. Both these forms of radiation are actually part of the electromagnetic spectrum (light) and are not a form of nuclear radiation like alpha or beta; however both are produced during any nuclear reaction and therefore should be mentioned.

Both gamma and x-ray radiation are sometimes referred to as penetrating radiation because they penetrate most substances.

Gamma radiation and x-rays are both highly energetic particles that are produced both naturally and by man-made devices.

Some examples of gamma radiation emitting substances are iodine 131, cesium 137, cobalt 60, radium to 26, and Technetium-99m.

X-rays can be produced electrically, as can microwaves and radio waves. Thick layers of dense substances such as lead or steal are necessary to shield against gamma rays and x-rays.

Now that we have discussed the various forms of radiation let's look at the amounts of energy released by a nuclear reaction, before we examine in detail how a nuclear power plant works.

The process of splitting an atom into two is known as fission. When a neutron splits an atom, an incredible amount of energy is released in the forms of heat and gamma radiation. The amount of energy released from the fission of a single U235 atom results from the fact that the products of the fission, e.g. newly created atoms and neutrons, weigh less than the original atom did before it was split.

The amount of energy released can be predicted by using the equation $E=MC^2$ or $E=M$ (as covered in the section of this book entitled "The Universal Equation").

A 100-Watt light bulb uses 100 Watt-hours of electricity, or 100 Watts per hour. The term "Watt hours" means Watts per hour.

The splitting of a single U235 atom releases approximately 88000 Watts of energy. To put this into perspective the splitting of a single U235 atom releases enough energy to power a 100-Watt light bulb for 880 hours (37 full days, 24 hours a day). That's a lot of energy. Now consider the fact that one pound of U235 is smaller than a baseball and contains billions of billions of atoms. The potential energy that can be harnessed from just 1 pound of uranium is enormous.

Most nuclear reactors have thousands of pounds of uranium fuel rods so they are capable of producing enormous amounts of energy for extended periods of time before it becomes necessary to change the fuel (re-fuel).

How a nuclear power plant works:

A nuclear power plant works almost like a conventional coal-fueled power plant, with the main difference being the source of heat that is used to drive the steam turbines which spin huge generators.

A nuclear power plant uses mildly enriched U238 as fuel (meaning 2% to 3% of U235 has been added to the uranium 238). The uranium is formed into pellets around the length and width of your baby finger. The pellets are arranged into long rods, which are then arranged together to form bundles of mildly enriched U238. These bundles are kept submerged in water.

Once the nuclear reaction is started it creates enormous amounts of heat that heats the water into steam, which then drives a steam turbine and finally a generator; however, if the reaction was left to run by itself, uncontrolled, the uranium would eventually get so hot that it would melt (this is called a supercritical reaction).

To prevent the above problem from occurring cadmium control rods (a material which absorbs neutrons) are lowered into the bundled uranium to slowdown the reaction so that it's just slightly supercritical.

Note: Another difference between nuclear and coal-fired power plants is that nuclear power plants use a secondary heat exchanger so that the water that comes in contact with the uranium, heats a radiator, which heats a second loop of water. This prevents radioactive material from coming in contact with the steam turbine, which prevents radioactive steam from being released into the environment.

Lowering the control rods all the way down into the bundled uranium will stop the nuclear reaction. Raising the control rods all the way out will eventually lead to a super critical nuclear reaction or melt down, like the one at the Chernobyl reactor in Pripyat Russia.

As a safety precaution 21 of the 50 control rods must always be left lowered into the uranium bundle. They are not lowered all the way in, but they're never raised all the way out. There is an operator who raises and lowers the control rods to control the temperature of the reaction.

The steam boiler or pressure vessel which houses the uranium is typically housed inside a large concrete liner which acts as a radiation shield; and that liner is housed within a much larger steel containment vessel.

Finally the steel containment vessel is protected by an outer concrete building which is supposedly strong enough to protect the steel containment vessel from things such as a crashing jet aircraft (jumbo jet).

Coal vs. nuclear power plants

A properly functioning nuclear power plant actually releases less radioactivity into the atmosphere in comparison to a coal-fired power plant. A coal-fired power plant also releases tons of carbon, sulfur, and other elements into the environment. That being said there are significant problems with nuclear power.

Historically mining and purifying uranium has not been a very safe or clean process. Also poorly designed or malfunctioning nuclear power plants can have lasting environmental consequences, such as the incidents at the Chernobyl plant in Russia, and at Three Mile Island in the United States, not to mention the Fukushima reactor disaster in Japan.

Transporting new and spent (used up) nuclear material poses some risk, also as of current there are no safe/permanent storage facilities for spent nuclear fuel.

These and other problems such as lenient government inspections, and governments being allowed to, and actually overruling safety inspectors and running unsafe reactors (which is clearly unacceptable for obvious reasons) have stirred public concern and are largely responsible for stopping the creation of new nuclear reactors.

Nuclear terms

Nuclear fission: The process in which an atom splits.

Nuclear decay: The process of an unstable element changing from its current form to a stable form. E.g. Hydrogen-3 is unstable and over time it will change to Helium-3, which is stable.

During nuclear decay (alpha, beta, or neutron) radiation is released. Alpha particles are positively charged and massive (often the actual nucleus of an atom). Beta particles are negatively charged and of low mass (beta particles have been shown to be electrons). Gamma rays are neutrally charged with no known mass.

Note: It is possible to make a naturally non-radioactive element become radioactive by bombarding it with neutron radiation. Alpha, beta, and neutron radiation are all subatomic particles, meaning that they are all smaller than an atom.

By now you're probably wondering how nuclear fission is induced. Nuclear fission is induced with particle accelerators (also known as atom smashers).

A particle accelerator works by electro magnetically accelerating particles (electrons) in a vacuum to speeds near the speed of light and then smashing them into an atom. If that atom is a U-235 atom then what you have is induced nuclear fission; however if it is a stable atom, such as an atom of phosphorus, what you have is light (light is nothing more then shattered electrons). Phosphorus atoms are much harder than electrons.

The CRT (cathode ray tube), the part of your TV that creates the picture that you see on your television screen is in fact a linear particle accelerator which accelerates electrons and smashes them onto a phosphorus coating on the back of the screen. In a color TV there are three beams of electrons and three different types of phosphorus used to create different colored pixels. The different levels of brightness amongst the various pixels are produced by using different amounts of electricity (different numbers of electrons).

Radiation particles are small enough that they can be held in an object's micro-gravity until they have broken down to such a point that they no longer are splitting off micro-gravity sized fragments of themselves, and are then pushed out of an object by the objects micro-gravity.

What is an atom?

An atom, in science, is the name given to a particle with a nucleus, made up of protons and neutrons, and a number of electrons orbiting this nucleus equal to the number of protons therein. Atoms can bond with each other to form molecules. E.g. A hydrogen atom and 2 oxygen atoms can bind together to form a water molecule.

An atom is the smallest piece of an element that is still that element. The element gold is simply many/lots of individual atoms of gold. Atoms are made up of protons (positive charge +), neutrons (no charge) and electrons (negative charge -).

Fusion and Fission

In nuclear fission an atom is split releasing energy. In nuclear fusion two or more atoms are joined to release energy; as an example our Sun combines two forms of hydrogen atoms (tritium and deuterium) together to form a helium atom, in this process one extra neutron and energy in the forms of heat, light, and radiation are released.

Scientists have been working to create a controllable nuclear fusion reaction for use in power plants for a long time, but thus far they have been unable to control the reaction in a confined space. Nuclear fusion creates less radioactive material than nuclear fission, and its supply of fuel is far more readily available and much safer to handle than uranium.

I hope this section provided you with an insight into, or a better understanding of, nuclear reactions and radiation.

In the next module we will be covering the very interesting topic "Dimensional Sciences." Thank you for taking the time to read this section, I hope you enjoyed it.

Dimensional Sciences

What is a dimension?

When we hear the term dimension we often think in terms of volume (length, width, and height); and most people believe that we live in a three-dimensional universe. This is true when speaking in terms of volume.

Some people believe that there is a two-dimensional world. Two dimensional like what an artist puts on paper, but even the lead from the pencil the picture was drawn with has length, width, and height.

The idea of something being two-dimensional is simply an idea. Everything we can touch or see in our universe is three-dimensional.

For the purpose of this module I would like to define the term dimension simply as a place of existence. For example everything everywhere that we as humans can touch and see is our dimension, and heaven would be another.

How many dimensions are there?

There are virtually an infinite number of dimensions. Everything has substance whether it exists in our dimension or not. The exact number of dimensions that exists is equal to the number of times in which one can be divided by two; and as I'm sure you are aware one can be divided in half indefinitely, just as easily as it can be doubled forever.

Our dimension is one of a possible infinite number of dimensions, with no beginning and no end.

How it all ties together. (Matter, Energy & Time)

As explained in the universal equation E=M, energy and matter are the same thing. One might call electricity or light energy but electricity is electrons, and light is shattered electrons (physical particles in motion).

Momentum is referred to as energy, as is heat, and both are examples of particles in motion.

That being said, all matter is always in motion. It really can't come to a complete stop. Some scientists will argue this with the theory of "Absolute Zero" meaning a temperature so cold that all atomic motion stops.

As mentioned in the section of this book entitled "The interactions between Gravity, Micro-gravity, and Ultra-micro-gravity" all things are kept together by gravitational forces. Also as I'm sure you have observed when something gets cold it becomes harder/ brittle (more micro-gravity is present because it is not being reduced by heat).

If it were possible to freeze an object to absolute zero everything the object was made of including the atoms and components of them would all come apart, because the force micro-gravity is comprised of extremely small particles in motion, and if the force was not effective due to a lack of motion of these particles (temperature), than there would be nothing left to hold everything together so it would immediately fall apart, as it falls apart what you have is particles in motion (temperature).

The point being that all particles of matter are always in motion, energy is defined as particles in motion. In that way energy and matter are both the same, not to mention they are both comprised of the same material, and as mentioned in the earlier module "Nuclear Reactions and Radiation" the energy released from nuclear fission is equal to the loss of mass (matter).

To recap a bit matter and energy are both particles in motion.

Now here's where things get a bit complicated and interesting.

Time:

Time is the motion of particles. Particles all move at different rates of speed and the comparison of these speeds is how we measure time. Here on Earth we have observed how long it takes for our planet to travel around the Sun. We then called this a day. We eventually further broke the day into a 24-hour period and came up with the measurement of an hour, minute, and second.

These are simply just measurements (broken into increments) of how long it takes for the Earth to rotate around the Sun.

This is our idea of time (particles in motion); it is simply just the way in which we humans currently measure time as it relates to us. In the end time is simply nothing more than particles in motion, and can be measured by comparison to the motion of other particles.

To recap a bit: Matter and energy are particles in motion and time is simply a comparison of the speeds of these particles: or as we humans have done, with a certain set of particles in motion at the same speed, which correspond to the rate our planet orbits the Sun, call it standard time and thereby relate all other particles in motion to this standard, calling it time.

Matter: Particles in motion.

Energy: Small fast moving, particles in motion.

Time: Particles in motion; measured by comparison to other particles in motion.

Now that you have an understanding of matter, energy and time; we will now be taking an advanced look at the matter in our dimension, and in other dimensions.

A dimension is simply a place of existence for matter. Matter is the motion of particles such as atoms and molecules (combinations of atoms). Gravity, micro-gravity, and ultra-micro-gravity are all forms of energy (lite weight, rapidly moving particles). All forms of gravity are released fragments of the particles of matter that they are holding together. Gravity particles accelerate away from matter due to dispersion. Particles of gravity do not have a speed limit in our dimension, and for other reasons do not stay in our dimension once spent.

Consider this when a particle of gravity has preformed its job (smashing another particle of gravity) it splits into smaller fragments. These fragments are no longer part of our dimension because they are so small (far, far, smaller than photons) that they can not interact with or affect anything in our dimension, in fact they cannot even be detected.

Dimensional Limitation

Anything that moves faster than the speed of light (187,000 miles per second) with the exception of gravity particles (which are far smaller and travel far faster) can not be considered part of our dimension since they can't be seen, and can not have any physical effect on matter or energy in our dimension.

Our spent gravity particles become matter in the next dimension. For simplicity purposes let's call the dimension obtaining matter from our spent gravity a lower dimension; and the dimension that fuels our dimension with matter from its spent gravity a higher dimension.

To clarify, when matter (spent gravity) is so small that it cannot interact with anything in our dimension; so small that it passes right through the nucleus of an atom without causing any interference or affect, then it is no longer part of our dimension.

All matter is made of tiny components, and although these components are small and mostly consist of empty space they make up everything in our dimension including us, our buildings, our planet, our solar system, our universe, our galaxy, and all time and space as we know it.

That being said there are also large life forms (presumably), structures, planets, etc. in the lower dimensions as well as in the upper dimensions.

For all we know there could be hundreds, thousands, millions, or even billions of planets orbiting right through our planet every second. It is not just conceivable, but actually extremely probable and likely, because all matter in all dimensions share the same space.

Planets in the lower dimension can orbit right through ours because the particles which they are made of (our spent gravity) are too small to have any affect on the matter in our dimension. Also the lower dimensions that are made of spent gravity from even lower dimensions are clearly even smaller, and again cannot affect matter in our dimension.

As for matter in the higher dimensions; the material our dimension is made of is too small to have any affect on the larger material which that matter is made up of in those dimensions. In fact spent gravity from the higher dimension just above ours, replenishes the matter in our dimension.

Our dimension looses matter permanently via spent gravity which becomes matter in the dimension directly below ours. However it constantly gains matter in the form of spent gravity from the higher dimension directly above ours. It's like a finely balanced matter replenishment system.

Think about this, since the material our dimension is made of (spent gravity from a higher dimension) is too small to interact with the material the higher dimension is made of; any beings/creatures in the higher dimension can be no more aware of us, than we are of them.

Dimensional Decay

There is a problem, the highest dimension will eventually dissolve (run out of matter); however, a new lower dimension will eventually form (the lowest).

Our dimension is among the lowest, as we recently had a big bang; but we are not the lowest, as the whole time our dimension has existed it has been continuously giving off matter (our spent gravity) which has certainly contributed to the creation of a new dimension, and possibly many new dimensions depending on how old our dimension actually is. No one knows for sure how many dimensions there actually are; we can take a guess, but no one really has any proof as to the exact number as of yet.

The Big Bang (How a universe starts in a dimension)

As the lowest dimension with physical matter ages it takes on new matter (spent gravity) from the dimension above it, and it throws off its spent gravity to the dimension below it (without matter).

The spent gravity particles cannot interact with the matter in the dimension above theirs (they are too small) and they do not interact with each other either, there are too few and they are too far apart.

They are however moving at very high speed (approximately several times the speed of light) and they are also continuously increasing in number as the spent gravity from the dimension above continuously adds to the volume of spent gravity in the dimension below it without matter. These spent gravity particles will eventually become matter after a series of random collisions that we will discuss further in a moment.

Essentially once the spent gravity particles in the lowest dimension, the one without matter, produce their own gravity then those spent gravity particles can be considered matter within that dimension.

The process in which free floating spent gravity particles from one dimension become matter in another is known as a big bang. For the purpose of explaining how this process happens and what it entails I will be using our universe as an example as we have recently had a big bang in our dimension which created our universe.

Note: There are likely many other universes in our dimension which are likely too far away for humans to detect with our current technology. That being said those universes are a tangible part of our dimension.

The Big Bang that started our universe in our dimension.

A long time ago our dimension had no actual matter (particles with their own gravity); just free floating spent gravity particles from the dimension above ours. At that point in time our dimension was the absolute lowest/newest.

As time went on there became more and more free floating rapidly moving spent gravity particles in our dimension from the dimension one level above ours (second newest) and from time to time 2 or more of them collided forming 8 to 12 particles of tangible matter with gravity. However because of the speed at which the initial 2 or more free floating particles were traveling at when they collided, the newly formed matter ("particles with gravity" e.g. atoms) quickly became separated by hundreds of thousands of miles of space.

That being said the occurrence of this matter creating process mentioned above began to happen more and more often as our dimension became fuller and fuller with free floating particles (spent gravity from the dimension just older than ours "second newest").

As time went on the volume of free floating particles (particles without gravity) and free floating atoms (matter with gravity) continued increasing beyond the point of inevitable collision, and eventually 2 atoms collided head on or nearly head on creating a tiny nuclear fission.

The fission would have remained tiny if it weren't for the fact that it was surrounded by free floating particles. The energy from the nuclear fission forced all the surrounding free floating particles into each other and away from the fission.

Many many free floating particles hit each other in such a small area and split into matter ("particles with gravity" e.g. atoms), and because there was, in an instant, so much matter in such a small area (the size of a two atom nuclear fission) it became extremely inevitable that many more rapidly moving free floating particles would immediately hit and smash each other creating many additional atoms; and that many more atoms would immediately collide, instantly leading to many more fissions over a much larger area and volume of matter (particles with gravity).

Once this occurred it was inevitable that it would occur again, and as it did the series of fissions were so great in number that they began to grow exponentially in number with each passing millisecond and could be considered non-stop.

The outward exponential amount of fissions continued occurring at incredible speeds until all the surrounding free floating particles had been converted into matter. This process of rapidly converting free floating particles into matter has been coined "The Big Bang".

Note: I mentioned fission but fusion also occurred during and after the big bang.

Fusion is the process in which two or more different atoms combine together to form new elements, and it is also the type of reaction which is constantly occurring on stars such as our Sun.

How a dimension ends (dissolves)

So far we have talked about how a dimension forms and where the matter comes from; now let's take a look at how a dimension ends.

I would describe a dimension ending as chaos over a long period of time, sort of similar to the urban legend regarding the US military's experiment named "Project Rainbow" A.K.A. "The Philadelphia Experiment", where matter which cannot exist in our universe, doesn't.

For those of you who are unfamiliar with the urban legend in regards to "Project Rainbow", the legend states that Einstein and other scientists in 1943 devised a way to make a boat (the warship Eldridge) invisible to magnetic mines via a degaussing magnet, but as a side affect the boat also became invisible visually as well.

Additionally the 1200 ton boat is said to have reappeared seconds later over 300 miles away. Moreover, some of the naval officers and crew on the ship are said to have become merged/fused to parts of the ship. Flesh and metal were bonded. Also some of the survivors were said to have disappeared and reappeared at a later time during a bar fight, and it is said that some of the survivors eventually spontaneously combusted, but again it is just an urban legend; however it may hold some truths and it does have some parallels with what theoretically happens when a dimension ends.

Let's say that you placed an object on a table, and then the size of the particles the object is made of were increased or decreased to the point that the particles which the object is made of could no longer interact with the particles our dimension is made of. If that were to happen then the object would indeed vanish. It would not just be invisible, it would in fact not be present in our dimension and not be able to interact with objects in our dimension, at least not completely; therefore there would be little if anything holding it to the table in our dimension where it was originally located.

There is a good chance it could come into contact with a stationary object in the dimension which it had entered, and be destroyed. Or the object may appear in space close to a planet or star and be drawn towards the gravity causing motion. Perhaps the object would exist in-between dimensions, where light/gravity from our dimension would have little or no affect on it.

There are millions of other imaginable scenarios, but in any case if we were to resize the particles which the object is made up of, it certainly would not be in the same location and very conceivably could be hundreds if not thousands of miles away from where it was, before we theoretically adjusted the size of the particles that it was made of. It is also possible that it could end up partially or fully inside some other object when it re-materialized (when we enlarged the particles it was made of). This is similar to the tale of the sailors' flesh being bonded with metal in The Philadelphia Experiment.

This to the best of my knowledge has not been done in real life, with the Philadelphia experiment, with magnets or the like, and to date no one has of yet conceived a way of doing this. There's really no benefit to doing it anyway.

To get back on track, when objects or microscopic parts of an object dissolve (permanently leave our dimension) they do not leave a visible hole, they simply do not exist. You can't have a hole in water, you can have a hole in the ground though, but you would not have the absence of everything (e.g. a magical void); the dimension which lost the object or part of the object would simply have a little less mass and get a little bit smaller.

Our dimension is constantly gaining and losing matter.

The highest dimension that exist (by the highest, I mean that it is made up of the largest particles) is constantly losing matter and not gaining any, so it is essentially getting smaller and smaller.

Another way of looking at the situation may be to say that the highest dimension is transforming into the dimension below it, and the one below it is constantly transforming into the one below it, etc.

One could say that all dimensions are in a constant state of transformation, with the lowest dimension being in a state of construction/creation (big bangs etc.) and the highest being in a state of deconstruction as it dissolves both via "particle by particle" and in large masses via black holes.

Black holes are areas of space, in all dimensions, where the particles the area is made of are running shy on material to maintain their size, and it doesn't take much to set off the conversion. By not much I mean something like a star going supernova (collapsing in on itself under its own gravity). That may sound like a major event, but in the grand scale of things it is a frequent and common occurrence.

How a black hole is formed

Consider this, all matter is made up of particles. In order for these particles to interact with one another they must be of a similar size. All particles are emitting smaller particles in the forms of gravity, micro-gravity, heat, light, magnetism, radiation, etc. and therefore are losing much of the material which allows them to maintain their size.

When a star dies (can no longer sustain fusion and fission) its massive amounts of gravity and micro-gravity cause it to crush itself. Although there is nearly the same amount of material/mass as just before it started to die, the amount of gravity producing particles increases as the pre-split gravity producing atoms slam into one another and split up further. Note: The atoms were already split during fission when the sun was active.

Once a half atom is split up 10 or so more times, what you have is a greater surface area to emit gravity, and therefore more gravity at all levels.

Once the dying star is effectively undetectable, because it is no longer emitting any sort of detectable emissions, at that point it can be considered a black hole.

A black hole emits so much gravity that any near by matter/objects; be it dust, an asteroid, a planet, or even a star, will be drawn by the gravity and push themselves towards and eventually into it.

The black hole contains such strong gravity that the objects which enter it will be crushed down to about $1/10^{th}$ of their original size or smaller, during this process both fusion and fission occur as the atoms the object are made of combine and split and split and split again, to form more gravity producing surface area; causing the black hole to gain gravitational strength.

Note: Very little, if any, light, heat, radiation, from the fusions and fissions are detectable as the black hole has too much gravity for these particles to escape. A black hole in one dimension is a huge source of free floating particles in the dimension below it. This surge of localized free floating particles in that said dimension would certainly contribute to a big bang.

In short, all matter in one dimension is made up of broken down matter (spent gravity) from another. There may be a top dimension and a bottom dimension, but there may also be an infinite number of dimensions because 1 can be multiplied or divided by 2 indefinitely.

No one really knows how many dimensions there are, or has enough data to make and educated guess. My personal belief on this is that there could be a near infinite number of dimensions.

I can't be certain as to whether or not, there is an absolute top or bottom dimension.

If there isn't an absolute top or bottom dimension, then there would be no dissolving of the top dimension and no creation of a new bottom dimension.

Either way because 1 can be divided by 2 indefinitely, there could be up to an infinite number of dimensions. Even if there was a top and bottom dimension (oldest and newest) there could still be an infinite number of dimensions. I believe there are billions and billions and billions of dimensions, a near infinite number, with a top and bottom dimension.

Either idea is feasible and neither can be proven right or wrong at this time.

I'm sure you found this module fascinating. In the next section we will be looking at the interactions between magnetism, heat, micro-gravity, electricity, and the affect of impact; which is both equally interesting and complex. I'm sure you will enjoy it as well.

The interactions between Magnetism, Heat, Micro-gravity, Electricity, and the affect of Impact.

In this section we will be discussing the interactions between magnetism, heat, micro-gravity, electricity and the affect of impact.

You may be wondering just what kind of complicated interactions could there be between these 3 forces and 2 forms of energy (heat & electricity). Well let's start out by looking at heat and magnetism.

Most of you probably already know that if you expose a magnet to a high source of heat for a prolonged period of time, the magnet will lose most if not all of its magnetic strength. Some of its strength will be lost permanently.

The opposite is true for cooling, the cooler you make a magnet (whether a permanent magnet or an electro-magnet) the more powerful it becomes.

This is because a magnet is always exposed to some heat; even room temperature could be considered a high temperature for a magnet. -275°C/-463°F is theoretically the coldest you could make a magnet, so while +20°C/70°F doesn't feel all that hot to you or me, when it comes to magnets +20°C/70°F could still be thought of as pretty hot.

Since heat causes some permanent loss of magnetic strength, why doesn't cooling it cause some permanent gain?

That's a good question. First off, even if you were to cool a magnet to -80°C/-112°F it still has some temperature. You could never really cool a magnet to absolute zero (-275°C), so essentially when you freeze a magnet your not really freezing it, your just lowering the temperature. However it is still of interest that high temperature causes partial or total loss of magnetic strength.

In order to heat an object to a temperature of 400°C enormous amounts of heat are required, and once an object has a high temperature like 400°C the object itself is actually emitting some heat. That is, some of the energy in the 400°C object is being released via temperature transfer, and some of the energy is escaping via heat, or even in the form of light.

Heat is a particle (as is light), and the flow of heat in or out of an object may drag (via flow) or push other particles out of the hot object; in the case of a magnet some of the particles being lost are "magnetic particles" (low gravity emitting particles which make up the force we know as magnetism, as discussed in the earlier module "Magnetism").

Heat doesn't rise, it moves evenly in all directions. As an example, when an airplane climbs the air temperature drops by 2 degrees Celsius for every 1000 feet of increased altitude. This is because as you gain altitude the air becomes thinner; hot air is thin, but not quite as thin. Hot air only rises because it is less dense than the surrounding air of a lower temperature.

This means, as a hot air balloon rises higher and higher, the air within the balloon needs to become hotter and hotter in order for it to continue to climb. However as the balloon is rising the surrounding air is becoming colder and colder as well as getting thinner and thinner. This temperature difference causes some cooling of the air within the hot air balloon.

Essentially heat moves evenly from its point of origin in all directions at rapid speed, and is different from temperature in that it can travel through a vacuum and through ice cold glass as well as some other materials, but not metal. Heat converts to temperature when it hits metal, and when the temperature of metal increases it emits some heat.

As an example of metal at a high temperature emitting heat let's take a look at a glowing oven element. A glowing oven element will emit heat (even for a short period after it has been turned off) which will travel through cold glass.

As an experiment you can run your oven on high with the door open until the electric elements are glowing. After you close the door, touch the glass and you will feel that it is cold, but move your hand away an inch or more and you will feel heat. Some of the heat from the hot elements is traveling through the glass and hitting your hand.

The glass is cold, but you can feel heat. This is because heat and temperature are two different things as mentioned elsewhere in this book. Heat is an energy that causes the vibration of particles such as molecules, atoms etc. Temperature is the speed at which these particles are vibrating. Everything emits some heat at all temperatures, even at temperatures well below the freezing point of water.

Here is another example of heat moving particles. For this example let's take a look at the highly misunderstood/under discussed phenomena condensation. I'm sure all of you know the general theory behind it. Most people believe that when a cool object is placed in a warm area, tiny droplets of water will form on it, which is true. E.g. A cold can of beer on a hot day will have tiny drops of water forming on it (condensation).

The reason this occurs is because there is more heat moving towards and into the can of beer than is leaving it (because the can of beer is colder than the air temperature), and it is the heat which pushes the water molecules onto the can of beer, where they are held in place by micro-gravity.

All of the excess heat (which we will discuss further in a moment) enters the can of beer, however most of the water molecules being pushed by the heat are unable to pass through the aluminum can, and therefore group together under their own micro-gravity and stick to the can as drips. Note: A very finite amount of water molecules are pushed into the can of beer via heat. Let's call this process molecular heat transfer or MHT.

Obviously both condensation and MHT work in reverse.

For example, If you were to take a can of pop out of a hot car and place it in a freezer condensation would form there because the heat being emitted by the hot can of pop would push the already existing water molecules in the air of the freezer onto its walls or onto other cold objects within it, where they would be held in place by micro-gravity. When enough water molecules accumulate together, what you have is the classic tiny visible droplets of water that most people know as condensation.

As for MHT, in the above example some of the water molecules the soda consist of will be pushed out of the aluminum can, they will travel directly through the aluminum. This occurs because there is more heat leaving the can, than is entering it.

Heat, water, and aluminum are all made of particles, as is everything. Typically a fluid won't travel through a solid, right? So how exactly does MHT work?

Molecular Heat Transfer (How it works.)

To explain how MHT works let's take a closer look at the 4 things present. Let's say we have an ice cold unopened can of soda in a warm room. The soda is the liquid, the aluminum can is the solid, and heat is the catalyst.

Heat is the particle/energy which pushes a finite amount of water molecules through the solid aluminum, and it can do this because it is affecting/pushing more than just the water molecules, it is also pushing/moving the aluminum molecules which the can is made of. This causes an increased vibration of the aluminum molecules and thus at certain random points in time larger spaces occur between some of them, occasionally large enough for a water molecule to travel through.

The aluminum molecules which the can consist of, although vibrating, are held in place by micro-gravity. The heat is not only increasing the vibration of the aluminum molecules as it hits them, but it's also pushing out a fair amount of micro-gravity which holds the aluminum molecules in place (in their variable fixed proximity from each other), thus increasing the vibration of the aluminum far more, which in turn increases the occurrence of a space between the aluminum molecules that is large enough for a water molecule to travel through.

For MHT to actually move a water molecule through the aluminum several conditions have to be met:

1) The heat has to cause a space between the aluminum molecules large enough for a water molecule to travel through.

2) In that very space/hole the heat must also push out some of the micro-gravity, which not only holds the aluminum molecules in place, but can also hold a water molecule as well, which would prevent it from passing through.

3) Once there is a microscopic hole in an area of the aluminum that has a reduced micro-gravity, there also needs to be a water molecule lined up directly with that hole.

4) The water molecule must be hit by enough particles of heat to push it through the hole, against the micro-gravity of other near by water molecules, which tends to keep water molecules grouped together. Or the water molecule and its micro-gravity both need to be hit at the same instant by the heat. (Micro-gravity and heat are both particles).

Note: The microscopic hole/space in the aluminum of the pop can was temporary. It only lasted a split second and was cause by changes in the proximity of the aluminum molecules.

That concludes my theory of MHT.

Let's take a moment to look at the relationships between heat, force of impact, micro-gravity, and molecular structures.

As discussed in an earlier section of this book entitled "The interactions between Gravity, Micro-gravity, and Ultra-micro-gravity", when one object slams into another the objects become half as strong (due to loss of micro-gravity) and as such it creates the appearance that as speed doubles force of impact increases by 4; however we know that force of impact only increases by 2 and the strength of the objects decreases by 2 (due to loss of micro-gravity), giving it the appearance that as speed doubles force of impact increases by 4. This is the main reason that force of impact has been so misunderstood for so long. Basically hard impact causes a partial temporary loss of micro-gravity, which is the force that keeps molecules in close proximity of each other/holds objects together.

Similarly heat also causes a partial and temporary loss of micro-gravity. As an example, if a car catches fire the roof might sag as it gets extremely hot and glows from the flames beneath it.

Essentially by the time the roof of the car has absorbed so much heat that it can no longer convert it to temperature, then at that point it has an extremely reduced micro-gravity holding the molecules that it is made of in place.

As the heat travels in and out of the roof, which already has an extreme temperature (rapidly vibrating molecules), it knocks out some of the micro-gravity particles being released by the molecules which hold them in place.

If there is enough excess heat the metal object, in this case the roof of a car, it may sag, melt, or change shape. This affect is temporary, once the fire is put out and the roof cools to the point that it is no longer being exposed to excess heat (So much heat that there is no longer sufficient available micro-gravity for an object to maintain its shape.), even if it still has a high temperature it will once again become solid.

That being said, although the loss of micro-gravity was temporary, the change of structure is permanent. Once the roof has cooled it will remain buckled.

Similarly, when you heat or impact a magnet, although the reduction of micro-gravity is temporary the loss of magnetic particles is permanent.

Sure you could re-magnetize it, but in that same way the roof of the burnt car could be straitened. Either way the alignments of the molecules have permanently changed, at least slightly.

On a side note, nuclear radiation is shattered atomic fragments which still produce micro-gravity. E.g. In a beta radiation emitting material the radiation being emitted is shattered atomic fragments which are temporarily holding themselves within the material via their own micro-gravity. Due to their rough and non-circular shape they rotate from the micro-gravity which they are releasing unevenly. The rapid circular rotation and collisions of the atomic fragments cause them to throw off smaller shards of themselves (radiation).

Note: Other than a fractional percentage of damaged atoms, something that becomes radioactive due to exposure to nuclear radiation is otherwise atomically/molecularly normal.

The Piezoelectric Effect

Let's take an in-depth look at the piezoelectric effect. Most of you have probably already heard of the piezoelectric effect and the bizarre scientific theories behind it. For those of you who haven't, a brief example of it might be an electronic BBQ lighter or electronic cigarette lighter in action (the kind without flint); when you push the button an electrical spark ignites fuel to produce a flame. The piezoelectric effect is the way in which the electric spark is created.

Here's how it works.

The piezoelectric effect is a combination of 3 forces:

1. Force of impact.

2. Micro-gravity (or loss of).

3. Electricity.

When you press down on the igniter button of an electronic lighter, a spring gets compressed fully and then released via simple plastic mechanical components. On the end of the spring there is a small, flat, circular piece of steel which is known as a hammer.

The hammer whips downwards and strikes a quarts crystal (essentially a type of rock), other materials are sometimes used. The quartz crystal which is mounted against a wire, releases a high voltage electrical current that travels through the wire and jumps to the electrode which is also the lighter's fuel valve. The fuel valve is grounded to the lighter's casing and the resulting spark ignites the fuel and creates the flame.

Where did the electricity come from, and where did it go afterwards?

The answer is actually quite simple and not surprising. Quartz is a material that has an excessively high micro-gravity and often it holds more electrons than materials that are surrounding it, which is the case in an electronic lighter.

When the hammer struck the quartz, the micro-gravity in the quartz was temporarily decreased. This caused the excess electrons in the quartz to push themselves out of the quartz, travel down the wire and jump through the air to the electrode; a fraction of a second later the remaining electrons changed direction and headed back to the quartz using the same path.

Some of the electrons became shattered (converted to light and heat) which created the visible spark. Additional electrons pushed themselves into the quartz from the body of the lighter. The body of the lighter received replacement electrons from whatever it was touching, for example your hand. This does not really affect you regardless of whether or not you are grounded to the Earth. If you were insulated from it than when you touch something with a grater amount of electrons than what is in your body, the electrons will flow back into your body, either slowly and unnoticeably or quickly (e.g. a static spark) depending on the amount of electrons and micro-gravity of the other object.

Why does it work like this? That's the simple part. As discussed in the earlier section of this book "Electricity and the generation thereof" we know that electrons (electricity) always flow from the greatest source of electrons to the greatest source of micro-gravity using the easiest path available.

Since the quartz has a greater amount of micro-gravity than the body of the lighter, it has more electrons. When the quartz is impacted by the hammer the micro-gravity is temporarily decreased so that it has a lower amount than that within the body of the lighter.

This causes the electrons to accelerate outwards from the quartz using the easiest path possible and creates the spark. When the micro-gravity returns to the quartz the electrons also will find paths to return to the quartz, either the original path, or if there are insufficient electrons to maintain the spark, than via an alternate path(s).

Note: In the above example the loss of micro-gravity was substantial and temporary, however it was not enough to cause any permanent changes to the quartz. Quartz is a material that has a natural excessively high amount of free micro-gravity, which is also the reason it is able to store more free/excess electrons than the materials surrounding it.

Let's look at another example of micro-gravity interacting with electricity. For this example let's say fresh clothes were just taken out of a dryer, and a sock sticks to a shirt rather than falling to the floor. Why?

Answer: Clothes taken out of a dryer can have an excessive amount of free/available micro-gravity. High amounts of micro-gravity can act stronger than gravity, for example a drip of water holding itself to the inside wall of an empty plastic water bottle.

Where did the excess micro-gravity come from?

When the clean wet clothes were taken from the washing machine and placed in the dryer, the clothes were saturated with water. Water has a naturally high amount of free micro-gravity, and as such it took some of the excess electrons from the clothes during the wash.

By applying heat the dryer is rapidly removing the water which has taken some of the free electrons from the clothes. The electrons leave with the evaporating water thereby causing the clothes to have a lesser than normal amount of free electrons, thus increasing the amount of available free micro-gravity that they have.

The high amount of available free micro-gravity within the articles of clothing causes them to push themselves towards each other and other objects, which creates the feeling we know as static cling.

Static cling can lead to a spark if something/someone with a greater amount of free electrons touches the clothes, because electrons flow from the greatest source of electrons to the greatest source of micro-gravity using the easiest path available.

Ultimately the clothes you have taken out of the dryer will appear to be attracted to other objects, whether clean or not, as long as the clothes has an excess amount of free/available micro-gravity, which they will until it receives electrons from something.

Static cling does not occur with clothes taken off a clothes line, as the clothes are dried slowly and are able to receive electrons from your house or the actual ground. In contrast the inside of a dryer is ungrounded as it is insulated from the physical ground via several materials such as rubber, plastic, concrete, wood etc. and it is also electrically isolated so that you cannot get a shock.

In a dryer the heat being blasted at the clothes temporarily decreases the available amount of micro-gravity within them, and therefore the clothing is able to hold far less electrons which is the reason they left with the evaporating water.

In essence, once the articles of clothing have cooled the available amount of free micro-gravity within them is substantially increased, since it is no longer being used to hold free electrons.

The clothing with the highest amount of free micro-gravity will draw electrons from the clothing with a lower amount of free micro-gravity, however even the clothing with the lower amount of free micro-gravity is resistant to giving up those electrons (it will give them up, but slowly).

Any free electrons in either the sock or the shirt are resistant to leaving, since both articles of clothing have excessive amounts of free micro-gravity. Until they receive electrons from something else, the excessive free micro-gravity will cause the articles of clothing to push themselves towards each other and other objects.

What is inductive heating and how does it work?

It wasn't until recently that I heard about inductive heating and I was briefly perplexed at how it could possibly work, but it does work!

I wanted to find out both how and why it works, however, by talking to most people it became quite clear to me that only about 1 in 100 people at best has ever even heard of inductive heating, so I will give an in-depth example below. Also please note that videos on inductive heating can be watched online at sites such as www.youtube.com

An inductive heating coil is a circular heating coil, which causes the temperature of the object within the coil to become higher than the temperature of the coil itself. For example if you had an inductive heating coil and placed a marble in the center of the coil (The marble is in no way actually touching the coil.) and plugged it in, the marble would glow bright red as the glass begins to melt at 1400 Celsius, yet the coil is far cooler at only 200 Celsius. Why?

Answer: When the coil reaches 200 Celsius in temperature it starts throwing off an immense amount of heat in all directions. The heat will of course be most concentrated at the center of the coil, since all the heat on the other side of the coil is dissipating as it is being thrown off into open space. In contrast all the heat on the inner part of the coil is concentrating at the center since it is being thrown into an area of limited space.

Also note that the temperature of an object within the center area of the coil becomes even higher because it is being bombarded by so much heat that it temporarily loses some of its micro-gravity. With a decreased micro-gravity the particles of heat that are hitting the molecules of glass cause them to vibrate even faster, therefore the glass reaches a much higher temperature.

Micro-gravity, Heat, and Water.

As an awesome example of heat affecting the micro-gravity of water, let's take a look at a coffee maker in action. As the coffee is being brewed, the water is being heated into steam which then recondenses as drips of water that run through a brew basket and filter containing coffee grinds. The coffee then drips into the coffee pot and is further heated by a heating plate directly below it.

This is necessary as once the water in the back of the coffee maker became steam, it quickly lost most of its heat and recondensed into only warm water, which would produce only luke-warm coffee so the heating plate is quite essential. This shows how the micro-gravity of a substance allows it to hold heat, once lost so is the heat.

As a further example let's say the pot of coffee was half brewed and we turned the coffee maker off, or unplugged it. There would still be enough heat in the heating coil to drudge up a little more steam, which would then recondense and fall into the coffee pot as drips of coffee, only they would certainly be cooler than the rest of the coffee in the pot.

As the cooler drips fall; the larger ones have so much weight (due to gravity) that they impact the lake of hot coffee in the pot with such force, that the micro-gravity of the coffee drip can no longer hold it together and its individual water molecules merge with the lake of coffee almost instantly. On the other hand, if the drip is small enough its micro-gravity would hold it together during impact, as gravity did not create enough force to cause it to split apart.

The result would be a nearly perfect sphere of coffee sitting right on top of the lake of coffee and it could continue to do this for several seconds since the micro-gravity holding the sphere together is stronger than the average micro-gravity in the lake, due to the fact that the lake of coffee has more heat. However, since the sphere of coffee is touching the lake of coffee the amount of heat within the sphere would quickly increase until it's nearly the same. At that point the amount of micro-gravity within the sphere would no longer be strong enough to hold it together against the combined micro-gravity of the lake of coffee and gravity of the Earth. Gravity would then cause the sphere to brake apart and merge itself with the lake.

Water or Air, which is heavier?

Most people believe that water is heavier than air, but this is truly not the case as I will explain using several examples.

First of all, a molecule of water is comprised of 2 hydrogen atoms and 1 oxygen atom. Hydrogen (a radioactive by product like helium) is one of the lightest substances there is, whereas oxygen is one of the heaviest.

The high micro-gravity of the oxygen atom causes 2 hydrogen atoms to push themselves towards it, when they combine the resulting combustion produces a water molecule (H^2O) and heat.

Hydrogen and oxygen are both found naturally in our environment. During the formation of water rapid heating occurred due to combustion. Afterwards the total amount of available free micro-gravity of each water molecule was highly decreased in comparison to the total micro-gravity of 3 atoms before they combined to create it.

This is because some of the micro-gravity from each of the atoms is being used/consumed as it holds them in place to maintain the water molecule as a molecule.

That being said water molecules still have a very high amount of available free micro-gravity.

The excess micro-gravity, unless temporarily reduced, causes the individual water molecules to push themselves towards each other to form liquid water. Liquid water contains more total mass than air because there are more water molecules occupying the same amount of physical space than there are air molecules.

Now let's look at a molecule of air (N^4O). A molecule of air is comprised of 4 nitrogen atoms, and 1 oxygen atom. Nitrogen is a larger and heavier atom than oxygen.

Essentially both air and water molecules have one oxygen atom, however a molecule of air has 4 Nitrogen atoms which are heavier than the 2 lighter hydrogen atoms of a water molecule.

You may be thinking if water is lighter than air, than why do we have lakes. The answer is proximity. Although a molecule of air weighs more than 3 times as much as a molecule of water, air molecules are spaced far further from each other than water molecules are when they are at the same temperature. This is because a molecule of air emits a lower amount of micro-gravity due to the close proximity of its nitrogen atoms, in comparison to a water molecule which emits a much higher amount of micro-gravity.

Let's say we fill a balloon with water, and fill a balloon of the same size with air. Although the individual water molecules weigh less than the individual air molecules, the water is in a liquid state whereas the air is in a gasses state; which means that the water balloon has 27 billion water molecules within it, whereas the balloon filled with air only has about 500 million. Although the air molecules weigh far more, there are less of them in a given space. When air is compressed into a liquid it weighs much more than liquid water, however this does not occur naturally in our environment.

The formation of clouds

As proof that water molecules individually weigh less than air molecules one only needs to lift his head and look at the clouds. Clouds are a mix of air molecules and water molecules. Clouds are created from moisture which is evaporated from lakes and wet soil or anything that is wet and exposed to heat.

When the Sun shines on moist ground, and most ground has some moisture in it, the heat reduces the micro-gravity of the surface water molecules, and some brake free from the water. Once free the individual water molecules float upwards indefinitely, whether exposed to heat or not, because they are lighter than air molecules. E.g. A cloud that formed during a cold winter day stays afloat at night when it's below freezing.

Eventually the water molecules come within a close enough proximity to each other to form water vapor (air and water); they are just close enough to each other to form a visible white cloud and to have a neutral buoyancy with the surrounding heavier air molecules.

A cloud emits enough micro-gravity that any free floating water molecules that rise up to it become a part of it and do not continue to rise (gain altitude). The more water molecules that become part of the cloud the stronger the micro-gravity that the overall cloud has and the larger it becomes; the larger it becomes the more water molecules it picks up.

Inevitably a cloud can only reach a maximum density ratio of water to air. As the air molecules are being squeezed out of the mix, the only thing keeping the water molecules from joining and producing drips of rain is heat from the Sun.

The heat temporarily reduces the micro-gravity of the individual water molecules (expanding there vibrational orbit), which allows air molecules to remain in-between them.

At night or when the heat is blocked (Some of the heat can even be blocked by the cloud itself if it is thick enough.) the micro-gravity of the water molecules causes them to push themselves together into drips. Once the water is in drip form it is far denser than air, and gravity causes the drips of water to push themselves towards the Earth.

Less available micro-gravity within newly formed materials

When water is formed by combining 2 hydrogen atoms with 1 oxygen atom there is less available free micro-gravity being emitted by the water molecule than by the three atoms alone. This is because the oxygen atom and the two hydrogen atoms are using a large amount of their micro-gravity to hold each other in place.

Let's say about 50% of each hydrogen atom, and 70% of the oxygen atoms micro-gravity is used. Although this seems like a lot, water still has a high amount of available free micro-gravity because the oxygen atom has more free micro-gravity than any other substance known to man; this is the reason that many materials will combine with it (combustion).

Note: The micro-gravity of a molecule works in the same way as gravity does on objects (groups of molecules). Please see the modules "Gravity. (What it is. How it works!)" and "The interactions between Gravity, Micro-gravity, and Ultra-micro-gravity" should you need more information.

Combustion

What causes combustion?

If you fill a balloon with $2/3^{ds}$ hydrogen and $1/3^{d}$ oxygen what you have is a balloon filled with $2/3^{ds}$ hydrogen and $1/3^{d}$ oxygen, not water. For specific reasons it takes a spark or source of heat to cause combustion, which will be covered in the example below.

For this example let's think of an engine, whether it be a car, lawnmower, snowmobile, chainsaw, or weed-whacker etc . . .

In order for the engine to run combustion must occur at the right time and this is controlled by the ignition system.

Picture the piston in an engine compressing the air and fuel mixture in the cylinder, once the tiny spark is delivered via the spark plug the air and fuel combine, not before.

This chain reaction occurs because the spark is a flow of electrons in which some of them are shattered to produce light including very high levels of infrared (heat); this heat causes a loss of micro-gravity and anti-micro-gravity, but more specifically a loss of balanced micro-gravity and anti-micro-gravity.

Picture the air and fuel molecules surrounding the spark. It is the side of the molecules facing the spark that experience the heat and loss of micro-gravity and anti-micro-gravity, but only on that side.

This causes the fuel and air molecules surrounding the spark to push themselves into it, and when this occurs they smash into each other and some combine to form new elements (exhaust) as well as heat.

The heat then propagates the reaction further until all of the molecules in the permitted time, have combined. In an engine this is usually not all of the air and fuel molecules, because they only have a certain amount of time to combine before the piston reaches the exhaust stroke and pushes the mixture out. Also the mixture is usually not perfect, and therefore there will normally be some unburned fuel (fuel molecules which did not combine with air molecules over the given period of time before the exhaust stroke).

Let's say that we had the balloon mentioned earlier, and that it had a near perfect mixture of 2 parts hydrogen and 1 part oxygen; then a static spark set off an explosion. Even though the blast would be amazing/immense, still not all the atoms would combine with each other to form water molecules because the blast itself would have heaved some of the atoms free. This is possible as the amount of heat created when most of the atoms combine, defeats enough of the micro-gravity on some of the remaining ones that they just float free and are pushed away from the explosion and therefore won't be able to combine with each other.

In essence no combustion is perfect, as the locations of the combustible atoms/molecules at any given point are predictively random.

On a side note if you were to take water and add electrons (electricity) to it, the resulting decrease in micro-gravity between the atoms of the water molecules would result in them separating back to their piece parts (hydrogen and oxygen atoms); which is no surprise since the energy (light and heat) created during the combustion that caused the atoms to join in the first place was nothing more than shattered electrons to begin with.

During the combustion when the heat temporarily decreased the micro-gravity of the atoms, the electrons they were holding were free to escape and many were likely shattered producing more heat thereby causing the reaction to propagate further.

As soon as the heat left the excessive free micro-gravity of the atoms (made even higher due the loss of electrons) then caused them to join (enter into a variable fixed proximity) with each other forming water molecules.

Simply providing a high surplus of electrons to the water molecules will cause them to revert back to their piece parts (atoms).

Laser Light

Have you ever pointed a laser on a wall or a coffee table and observed the light. If so you will know what I'm about to speak of, if not feel free to try it either before or after you read what I am about to tell you.

Let's say you shine a laser pointer on a coffee table so that the dot is about 2 feet away from you. Instead of looking directly at the red dot, look near it and you will see a very unique phenomenon.

You will see thousands of tiny black dots appear and immediately explode releasing red (laser) and yellow (visible spectrum) light. What is interesting about this is the black dots appear even during daylight and are so dark that they cannot be seen through. This is because they are absorbing all light, both visible and laser.

You may be wondering what these black dots are and why they are produced by laser light? Laser light has a higher amount of micro-gravity than visible light which is why a laser beam's path doesn't widen much over large distances. It does widen, just not as much.

When you point the laser on a coffee table, through a piece of white rubber, or through other translucent substances you will see the black dots.

These dots occur because when the laser light hits the coffee table most of the light is absorbed into it and converted into temperature. The micro-gravity is not absorbed into the coffee table and builds up around the area where the laser light was absorbed. This build up of micro-gravity is known as grouped micro-gravity. Grouped micro-gravity is micro-gravity that is all by itself (without tangible substance), held together by releasing tiny parts of itself (ultra-micro-gravity).

Grouped micro-gravity absorbs all forms of light and heat which is what creates the black dots you see from the laser. The light and heat the grouped micro-gravity absorbs, defeats the very grouped micro-gravity that absorbed it. This is because the light and heat release their own micro-gravity, which erodes away at the amount of available grouped micro-gravity (using it up).

Finally once the grouped micro-gravity is used up (caused by the lights own emission of micro-gravity) the light and heat escapes and accelerates away from each other. This is called a grouped micro-gravity explosion; the grouped micro-gravity itself hasn't exploded but rather has been used up; so that the light and heat are free to explode away from each other in all directions.

Laser light produces more micro-gravity than visible light does, which is one of the things that make it unique.

Cutting lasers

A cutting laser utilizes a higher number of photons than a laser pointer. It has a more concentrated beam.

When a cutting laser is cutting metal two things are happening.

Firstly the photons of the laser light are smashed to pieces as they hit the metal, this greatly increases the temperature of the metal as the photons vibrate the molecules during impact.

Secondly the grouped micro-gravity from the smashed photons builds up sufficiently to defeat the micro-gravity of the metal molecules causing them to leave their home vibrational positions. Once free they are able to combine with other substances in the air, or with the air itself.

As the metal molecules are removed from the metal, the two sides of the metal become separated/cut.

The smashed photons from the cutting laser will inevitably become so small that they will no longer interact with anything in our dimension, and the material which they were comprised of will eventually become matter in the next dimension, as discussed in the module "Dimensional Sciences."

Microwaves

Microwaves work by agitating molecules (especially water molecules); this is accomplished by fracturing electrons. These fractured electrons (microwave radiation) act like heat in the way that they smack into molecules causing them to vibrate, which increases the temperature (molecular vibration) of an object being microwaved.

That being said there is one main difference between heat and microwave radiation which I will explain using this analogy. Heat is like fine ground pepper, ground nearly into a dust, whereas microwave radiation is like taking a pepper corn and splitting it into 3 or 4 pieces.

On a sub atomic level microwave radiation is fragments of electrons that have been split into 3 or 4 pieces, as opposed to heat which are the pieces of an electron that have been split into millions of pieces. Both heat and microwave radiation have more than enough mass to vibrate molecules via impact, however there is a difference between them as I am about to explain below.

Let's say we have a group of 1,000,000 molecules at 60 degrees Fahrenheit and we are heating them to 90 degrees Fahrenheit using microwave radiation to increase the temperature. In this example only 1000-1200 units of microwave radiation hit the 1,000,000 molecules per second for a duration of 30 seconds. So in total 30,000-36,000 units of microwave radiation hit 1,000,000 molecules over 30 seconds. Those units of microwave radiation then split into many smaller fragments more closely resembling heat, and those fragments then go on to hit other molecules causing them to vibrate (causing increased temperature).

Units of microwave radiation will, like electrons, move towards the greatest source of micro-gravity using the easiest path available. The greatest source of micro-gravity in microwaved food is water.

Another excellent source of micro-gravity is metal. For example if you microwave a CD you will see light which greatly resembles sparks as the units of microwave radiation accelerate towards the metal part of the CD and hit it, or hit each other once within the metal and split apart.

Since microwave radiation is comprised of electron fragments, further splitting them will create heat and visible light.

Note: Metal has more micro-gravity than water; this includes liquid metals such as mercury.

Microwaving water will not produce visible light, as water is a liquid which means the molecules of water are already vibrating rapidly and the impact of the units of microwave radiation is softer (more cushioned) or more abrupt and much harsher. In either event the fragments do not split to the size that we can see as visible light. They are either much larger (a form of heat) or much smaller (a form of micro-gravity).

Essentially the units of microwave radiation hit water molecules and cause them to vibrate, during impact the units of microwave radiation split up further into fragments similar to the size of heat, which go on to hit other water molecules causing them to vibrate, and then split up further becoming closer to the size of micro-gravity causing a temporary decrease in available micro-gravity within the water which allows the water molecules to vibrate even more violently (higher temperature).

Example 2: If you were to microwave an incandescent light bulb it would actually light up as if it was being powered by electricity. A clear one might also produce other visible colors such as purple or green streaks within the gasses of the bulb, however in less than 15 seconds it would get so hot that the glass could split apart.

In this example the units of microwave radiation are acting like electrons in that they are traveling to the greatest source of micro-gravity (the metal in the bulb), however once they impact with the metal in the bulb they split up into heat, light, and micro-gravity.

The increased micro-gravity along various random points in the filament and in the connecting metal causes some of the units of microwave radiation to travel within the molecules of the filament and in many cases they collide with other units of microwave radiation either within the filament or heading towards the filament. This produces immense heat and visible light, as well as high temperatures in some places. The inconsistent and uneven high temperatures can cause the glass to crack or brake.

In summery splitting electrons produces heat, light, and increased micro-gravity. Once the micro-gravity is spent, it becomes matter in the next dimension as discussed in the previous section of this book entitled "Dimensional Sciences."

We will be looking further into the losses and gains of energy in the next module entitled "Free Energy." Please read on and enjoy.

Free Energy

Free energy is a concept that has been around for centuries. Essentially the idea is to get energy with out expending any sort of fuel or energy input. For as far back as I can remember people have been trying to make devices that could produce free energy. Some examples are gravity wheels, and various forms of magnetic motors that do not use electricity. None of these devices have been successful thus far.

Does free energy actually exist?

The closest thing we have to free energy is solar panels and wind generators, however one could argue that they are not sources of free energy as the solar panel is using energy from the Sun and the wind generator is using the momentous energy of the wind.

Think of this, even if so called free energy devices actually worked, they would by default be breaking their own definition. If the gravity wheel worked it would be using gravity as a source of input energy (fuel) to drive it. If the magnetic motor worked it would be using magnetism as a source of input energy (fuel). Therefore there really is no such thing as free energy, or is there?

I would argue that all energy is free energy in that energy is constantly being gained and lost from within our dimension. For those of you who are skeptical of this let's take a look at a different example of free energy as follows below.

<u>*Example of Free Energy*</u>

We've all heard of potential energy right? An example of this is a car at the top of a hill, if someone were to release the hand brake it would roll to the bottom of the hill, it has the potential to move (potential energy).

Another example of potential energy would be if you were to press down on a spring with your fingers, because if you were to release your fingers it would expand/move. It has the potential to move/expand as do the springs in the suspension of your car. I'm sure you would agree that they have potential energy.

Here's the thing, there is no such thing as potential energy. For the car at the top of the hill with the parking brake on, gravity is causing it to try to push itself towards the bottom of the hill with a certain amount of force (based on its mass); regardless of whether or not it is being restrained by the parking brake. Even if the car is restrained by the parking brake and is not actually moving, the force of gravity is there and it is continuous, regardless of whether or not there was even a hill.

Let's say you take a spring and compress it between your thumb and your index finger, you might say it has potential energy because it will expand if you release your finger. I say it will continuously try to expand whether or not you release it. It will continuously apply force against your finger muscles the whole time it is compressed, that continuous force is energy. Would you call that force potential energy or free energy?

As for the above, a person may argue that while a spring is compress between one's finger and thumb it has potential energy because if released it will expand. That person might also argue that just because it is constantly trying (applying force) to expand, until it does it hasn't released energy. I would argue that the force the compressed spring is applying, whether allowed to expand or not, is in fact energy.

If you were to try and push a bus, and you pushed as hard as you could but it did not move, would you say that because the bus did not move you did not expel force and expend energy. When you tried to move the bus you were expelling force the whole time you pushed on it, and in doing so you were certainly expending energy; the bus did not move because it would take more force to move it than was provided.

If you were to compress a spring with your finger and thumb, yes there is potential energy, but there is also the constant energy of force. Even though you are applying so much force that the spring cannot expand, the spring is still expelling a constant force to try to expand, even though it cannot until you let it.

If you were to compress a spring with a force of 1 lb.; once compressed you would need to continue to apply a constant force of 1 lb. to keep it compressed because it is applying a constant 1 lb. force to expand.

While holding the spring compressed with a force of 1 lb., your body is expending electrical energy via your nerves. It is using fuel energy in the forms of oxygen and sugars. Your muscles (despite the fact they are not moving) are producing heat as a by product of the energy which they are expending in order to apply 1 lb. of force against the little spring. Would you call the force that the 1 lb. spring is applying against the strength of your muscles free energy? Well I would, but probably not for the same reasons as you.

You may argue that the constant force the spring is applying against your muscles is free energy, after all the spring is not burning any fuel, or making any heat. I could somewhat agree with that after all the spring is not consuming any fuel or other input energy, however there is energy coming from somewhere isn't there?

So where is the energy coming from?

The energy is coming from the micro-gravity within the spring. The reason I would consider this to be free energy, is because the micro-gravity is being spent at the same rate whether the spring is compressed or not.

The molecules in the spring are aligned when it is in the expanded position, compress it, and micro-gravity causes the molecules to apply force to try to push themselves back to their original positions. If they are prevented from actually moving back into their original positions they will keep trying for some time, but not forever.

Eventually the molecules will push themselves into new positions. As an example if a spring is left compressed for a really long time (years) when finally allowed to expand, it may not expand all the way or with the same amount of force as before since some of the molecules in the spring will have changed positions.

Why do molecules in a spring shift position if the spring is left compressed?

The molecules in a spring, or any other object for that matter, are being pushed towards each other by micro-gravity. Micro-gravity as you know works like gravity but on a smaller scale. The molecules in the spring are of the same size as they are made up of the same material, so each molecule is emitting roughly the same amount of micro-gravity.

Each time that the micro-gravity of the compressed spring becomes reduced by heat, the heat will push some of the molecules into new positions.

Micro-gravity is causing the molecules of the compressed spring to continuously apply force to try to return to their original positions, despite the fact that it is impossible for them to do so while the spring is compressed.

If the micro-gravity is unevenly reduced (by heat) some of the molecules will shift into more balanced positions. They will be pushed into these new positions by the heat. The new positions will effectively be the same or similar to that of what they were in when the spring was compressed. Therefore if allowed to expand, it will apply less force to do so, or it might not expand as much.

The reason why molecules cannot physically touch each other

Whether it's two molecules, two atoms, or two planets of the same density (mass and size); neither one will touch based solely on its own gravitational force. Sure gravity causes the two planets to push themselves towards each other, as does micro-gravity with two molecules or two atoms of similar density.

That being said there is more at play between two planets or two molecules of the same density than just gravity or micro-gravity. Planets emit both gravity and anti-gravity, as do all objects. Molecules & atoms emit micro-gravity and anti-micro-gravity, as well as gravity and anti-gravity which are also, in a way, all forms of free energy.

Consider this, if the objects in question whether planets or molecules are heading towards each other at a high velocity than they could collide, as their momentum (mass * speed) would undoubtedly exceed the objects anti-gravity or micro-anti-gravity.

Two molecules in close proximity pushing themselves towards each other are restrained from touching due to anti-micro-gravity.

Two planets heading towards each other at a relatively slow velocity would also be unable to physically touch/collide as gravity will only cause them to push themselves so close to each other, after which anti-gravity will keep them separated.

This means that the two planets or the two molecules respectively, would eventually settle into a variable fixed proximity from each other. They might be pushed closer together or further apart from each other at times via other forces or impacts; but gravity/ micro-gravity is working at full strength to cause the planets/ molecules to push themselves towards each other, and the only real thing that is preventing them from getting to close is anti-gravity/ anti-micro-gravity respectively.

In the case of two planets with an established variable fixed proximity it would take a force greater than the force of the combined two planets gravity to separate them from their variable fixed proximity. It would take equally the same amount of force to push them together and make them collide.

It's hard to imagine such forces, but they do exist; for example two huge meteorites traveling at 100'000s of miles per second could certainly cause the two aforementioned planets to impact each other by hitting their opposing surfaces.

Two meteors traveling at high speed hitting each planets facing surface could separate them. Keep in mind those types of forces would ultimately damage or destroy the planets.

The same thing applies to molecules, only instead their variable fixed proximities are established and maintained by micro-gravity and anti-micro-gravity, whereas with the planets it was established and maintained by gravity and anti-gravity.

A molecular example of this is molecules in a variable fixed proximity. The variable fixed proximity could be temporarily changed by heat as it impacts the molecules.

The proximity could be permanently altered and some molecules destroyed by:

1) So much heat that molecules continuously impact each other and other molecules. E.g. An aluminum alloy pop can in a camp fire.

2) Extreme impact with each other. An example could be steel molecules on the hood of a car in variable fixed proximities to each other. Some will smack into each other and brake up during an extreme impact from outside forces, such as what you would have during a car accident. On some parts of the car's hood the variable fixed proximities of the molecules it is made of will become altered, e.g. creased or bent metal.

Anti-gravity & Anti-micro-gravity

What are these forces?

Anti-micro-gravity keeps two or more atoms or molecules of equal or similar size and mass (density) from actually touching each other, despite the fact that micro-gravity is causing them to push themselves towards each other. Micro-gravity pushes the atoms or molecules as close together as it can, but they can not touch as anti-micro-gravity pushes them apart once they are in a close enough proximity.

This creates a variable fixed proximity between the two atoms or molecules.

Anti-gravity does the same thing as anti-micro-gravity but on a much larger scale, affecting planets, stars, and large stellar objects.

Anti-gravity and anti-micro-gravity are present where more than one particle/object are in a variable fixed proximity, for example anti-gravity is helping to keep our solar system's intricate orbital parameters in check.

For instance it is helping by pushing the Moon away from the Earth while gravity is causing the Moon to push itself towards the Earth; however, because of the difference in size and mass (density) between the Moon and the Earth; if the two bodies were to become completely stationary, gravity would eventually push them much closer together than currently possible (due to no centrifugal force being present).

That being said there may not be enough anti-gravity generated between them to overcome the force of gravity. Keep in mind without external forces being present, e.g. meteorites, than there wouldn't be enough energy to get them close enough together to hit each other anyway, regardless of the amount of anti-gravity generated between them.

Certainly if the Earth and another planet of equal or similar size and mass were placed close enough together in space without orbiting each other, gravity would be unable to push them together against the force of anti-gravity.

The planets would accelerate towards each other and then slow to nearly a stop. A slow orbit would result between them from the momentum caused by the gravity pushing them towards each other. That being said they would not touch regardless of the gravity. Instead they would end up in a variable fixed proximity to each other, due to the strength/amount of anti-gravity between them, and this is proportional to the planets in question being of the same or similar size and mass.

You're body is emitting anti-gravity against the gravity of the Earth, however you do not have an equal or similar density to the Earth, therefore the amount that your body emits is negligible and not something you ought to be aware of.

Atoms release micro-gravity, gravity, and radiation (in the forms of alpha and beta particles), as part of their natural atomic decay. That is, atoms release physical parts of themselves including gravity, micro-gravity, alpha, and beta particles (radiation); and overtime (hundreds to millions of years) become smaller (become different atoms) for example a carbon-14 atom emits radiation, micro-gravity, and gravity and transforms to a nitrogen-14 atom. A hydrogen-3 atom undergoing beta decay becomes a helium-3 atom. Note: Sometimes an atom of one substance will split in half and become two atoms of a different substance, this is known as spontaneous fission.

How are the forces Micro-gravity and Anti-micro-gravity created?

In the following example I will attempt to explain how anti-micro-gravity is created. Let's say we are observing two molecules of plastic, which are part of a plastic item such as a cell phone. Imagine that we can see the molecules vibrating in their variable fixed proximities. The molecules are emitting micro-gravity in all directions which is causing them to push themselves towards each other, but when they get very close they repel from each other due to anti-micro-gravity.

Picture each of the two molecules releasing hundreds of billions of tiny particles in all directions at the same time (micro-gravity).

Since all the particles are being thrown away from the molecules in all directions; neither of them has any reason to move, but as the molecules get closer to each other, some of the tiny particles being thrown off the first molecule brake some of the tiny particles on the second molecule before they can be thrown off, and some of the tiny particles on the second molecule brake some of the tiny particles on the first molecule before they can be thrown off.

This causes a reduced force in-between the molecules, so the two molecules have more force pushing them together than in any other direction, regardless of their orientation.

If it wasn't for anti-micro-gravity preventing the two molecules from moving closer to each other indefinitely, than they would surely collide.

As the molecules get closer together more of the tiny particles from the first molecule smash more of the tiny particles on the second molecule before they can be thrown off towards the first molecule and vice versa. This causes an even grater decrease in the force between the two molecules, so they want to push themselves together even harder. The closer they get to each other, the grater the affect micro-gravity has on the two molecules.

Once the molecules get really close to each other, perhaps a distance of approximately 4-7 molecules in length apart, the following will occur frequently.

One or more of the micro-gravity particles from each of the molecules will collide with each other in-between the molecules, rather than smashing one another on the surface of the adjacent molecule or being thrown off into open space.

This collision fractures the two or more micro-gravity particles into tens of thousands of pieces (anti-micro-gravity particles), and these pieces explode away from the point of collision in all directions; about 50% to 70% of them then hit the source molecules (the molecules which were the original source of the micro-gravity particles that collided). These impacts push the molecules away from each other.

Note: For the purpose of simplicity, I talked about "two or more" micro-gravity particles colliding in-between the source molecules, in actuality there would be millions of micro-gravity particles colliding in-between the source molecules at around roughly the same time.

When the molecules are at a distance of 3-4 molecules in length apart, it is more likely that the micro-gravity particles impact each other in-between the molecules with a greater frequency than impacting the micro-gravity particles on the opposing/adjacent molecule's surfaces before they otherwise could be thrown off. The impacts in-between the molecules cause the micro-gravity particles to split up creating anti-micro-gravity particles.

The closer the molecules get to each other the more likely micro-gravity particle collisions in-between the molecules will occur, and the more anti-micro-gravity will be created causing the molecules to be pushed apart.

Micro-gravity is causing the molecules to push themselves towards one another, when they get close enough anti-micro-gravity pushes the molecules away from each other with more force than the micro-gravity creates; this is what causes the variable fixed proximity between the molecules.

The example I just demonstrated talked about two molecules for simplicity, however the forces micro-gravity and anti-micro-gravity apply to every molecule in an object, as well as every atom that each molecule is made up of. Each molecule (except for the surface molecules) is completely surrounded by other molecules of the same type.

Try to picture a molecule that is completely surrounded by a sphere of other molecules, all of which are in variable fixed proximities to the center molecule that they are surrounding. Each of the surrounding molecules are also being completely surrounded by their own sphere of other molecules in variable fixed proximities as well, etc.

All of these spheres of molecules are interconnected in that each of the spheres is made up of some of the same molecules as some of the other spheres. No matter which molecule you look at, with the exception of the surface molecules, it is surrounded by a sphere of other molecules in variable fixed proximities to it because of the interactions caused by the forces micro-gravity and anti-micro-gravity.

Note: The surface molecules have closer variable fixed proximities with the molecules below them, which form half spheres around them.

How are the forces Gravity and Anti-gravity created?

Gravity particles and micro-gravity particles are released from atoms as part of their natural atomic decay. Note: Gravity particles are much smaller than micro-gravity particles, but there are many more of them.

Anti-gravity is created in the same fashion as anti-micro-gravity. When two stellar objects of equal or similar size and mass get close enough to each other, anti-gravity is created. It is a byproduct of gravity particles that collide and shatter in-between the planets. This is why if you were to put two planets side by side in space, in a close proximity; they would never touch each other.

Gravity and Anti-gravity in perspective

If two objects of similar size and mass fail to emit much gravity due to their small size, then very little anti-gravity is created. For example let's say two astronauts/cosmonauts left the international space station, and using jet packs positioned themselves 1 meter (3.3 feet) apart. The astronauts/cosmonauts are stationary to each other.

They are both emitting gravity and consequently anti-gravity is being created; however, with the amount of gravity they are emitting it would take years for it to cause them to move closer to each other. Due to the slow momentum (years to travel two feet) their anti-gravity would be sufficient to keep them apart. If two objects were of considerably different size or density than that may not be the case.

That being said if someone else left the international space station and threw one of the two astronauts/cosmonauts a baseball and that person caught it; that would be more than enough force to cause the two astronauts/cosmonauts to impact each other, or move away from each other, depending on who caught it.

In conclusion generally people think of gravity/anti-gravity as great forces, but that's only because we are constantly experiencing gravity from a much larger object (the Earth) with a greater mass and density than ourselves.

The Science

The science of theoretical physics is one of grate speculation and hypotheses which rely on proven facts, the results of new and not publicly available experiments, as well as both direct and indirect evidence.

This book is progressive and opinional, and may contain certain things which might not be completely true; however, I'm sure that it contains many things which are true. If it has even made you question or think about one piece of science (whether theory or fact), than it has done its job.

I sincerely hope you question the things about science which you think you already know; such as the things you have been taught in school, or have seen on television.

Even though many people and organizations present theories as fact (e.g. magnetism), many of the theories that you have seen thus far are based solely on speculation, not indirect evidence.

This book contains a lot of indirect evidence, but indirect evidence is evidence none the less. Einstein and many other scientists used indirect evidence to support their theories; most were later proven to be at least somewhat true.

This concludes my view of the intricacies of our dimension at this time.

I hope you found this book educational, enlightening, or at least somewhat interesting; moreover I hope you enjoyed reading my many theories and beliefs on how and why the invisible forces that are all around us do what they do.

About The Author

Born in 1978 Justin Klickermann has attended five separate schools, and has worked for nine different companies. Justin has a strong technical and business background, and has enjoyed careers in technical support, sales, and customer service. Justin has also worked in the airline industry, food chains, security industry, and he even had a job in shipping and receiving.

Justin is multi-talented and has vast interest ranging from sports, music, and exercise to science, physiology, and psychology.

Justin's actual lifestyle, creativeness, powerful self-motivation, strong dedication to achieving his goals, and careful attention to detail, all play a role in his writings, as does the support of his many friends and loving family.

Other Books By Justin Klickermann

- About The Author (paper back & ebook)
- The Secret Spectrum (ebook)
- Harry's Little Helpers (ebook)
- A Harsh Reality (ebook)

About The Author

About The Author is a murder mystery story in which the main character Justin Blackman, steals an unpublished murder mystery novel from his recently deceased roommate.

It is a first-hand account of how Justin Blackman steals the novel, hides the evidence, gets the novel published, and makes over a million dollars doing it.

The problem is that other people during the story begin to surface who know that the novel Justin Blackman has claimed authorship of, was actually written by his dead roommate Fred Banoner.

Justin Blackman must kill everyone who knows about it, and anyone who finds out, while evading police, and destroying all the evidence linking him to these murders; in order to keep his secret, reputation and money.

Will he be successful?

The Secret Spectrum

There is an 8th spectrum, a spectrum of light, which is unknown to man. There are invisible plants, animals and even creatures similar to Humans, they can see us, but we can't see them. What happens when we meet? Does this explain ghost/poltergeists?

Do these creatures have body heat just as we do? Do these creatures help magicians who do real magic tricks? Who are these creatures? How can you see one? What if the creatures in the secret spectrum were like chameleons, could they reveal themselves?

All this and more will be thoroughly covered and fully explained in the story "The Secret Spectrum"

Harry's Little Helpers

Harry's Little Helpers is a twisted, dark, and disturbing story in which the main character abuses prescription drugs and alcohol. He is generally depressed and dissatisfied with his life.

One day he discovers the presence of two aliens. The aliens make a deal with him, which greatly increases the quality, and extends the length, of his life; but there's a cost!

(Warning: This book is not intended for the mentally ill. Do not read if you have Schizophrenia, or have a history of depression or mental illness.)

A Harsh Reality

A Harsh Reality is a fictional story portraying the interactions between sociopaths, psychopaths, and the average person. It is a story of betrayal, deception, manipulation, and discreet reve